ABHANDLUNGEN
DER AKADEMIE DER WISSENSCHAFTEN DER DDR
Abteilung Mathematik-Naturwissenschaften-Technik

Jahrgang 1977 · Nr. N 9

Environmental Mutagens

Proceedings of the Sixth Annual Meeting
of the
European Environmental Mutagen Society
organized under the auspices of the
Akademie der Wissenschaften der DDR
by the
Zentralinstitut für Genetik und
Kulturpflanzenforschung der AdW der DDR — Gatersleben

GERNRODE (GDR) September 27 — October 1, 1976

Edited by: H. Böhme
J. Schöneich

AKADEMIE-VERLAG · BERLIN · 1977

Erschienen im Akademie-Verlag, 108 Berlin, Leipziger Straße 3–4

Lizenznummer: 202 · 100/278/77
Gesamtherstellung: VEB Druckhaus Köthen, DDR-437 Köthen
Bestellnummer: 762 466 2 (2001/9/77/N) · LSV 1325
Printed in the GDR
DDR 26,– M

Preface

The 'Sixth Annual Meeting of the European Environmental Mutagen Society' was held in the German Democratic Republic from September 27th to October 1st at Gernrode/Harz.

During the preparation of the Meeting a new structure with respect to the scientific program was introduced. The purpose was to concentrate on selected topics and to provide possibilities for extensive discussions and personal talks. Three symposia with invited lectures on topics of present and future interest were arranged:

Symposium I dealt with the importent problems involved in modifications by endogenous and exogenous factors of the mutagenic action of chemical agents.

Symposium II was entitled "Dose effect relationships, threshold effects, and the question of chronic versus acute exposure".

Symposium III concerned problems involved in studies on the effects of mutagenic chemicals in risk populations. This topic was of special interest with respect to the Seveso calamity.

The "Short Communication Session" was directed to "Recent methodological advances in mutagenicity testing in mammals".

A "Poster Session" was arranged which concerned studies in connection with the topics of the symposia, the short communications and on all other aspects of environmental mutagenesis.

Two panel discussions took place: "Problems with bacterial screening tests incorporating metabolic activations" and "Testing chemical compounds in a multicomponent test system".

Abstracts of all reports will be published in 'Mutation Research'. The present "Abhandlungen der Akademie der Wissenschaften der DDR" contain the lectures given during the symposia, the opening address given by B. A. Bridges (UK) and the closing lecture given by F. J. de Serres (USA).

We are grateful to the Academy of Sciences of the GDR for sponsoring the meeting and to the Akademie-Verlag for its help in providing quick publication facilities for these proceedings.

H. Böhme
J. Schöneich

Contents

Lectures

Symposium I

Modifying effects in chemical mutagenesis

Symposium II

Dose-effect relationships, threshold effects, and the question of chronic versus acute exposure

Symposium III

Population exposure and genetic monitoring

Lectures

Recent Advances in Basic Mutation Research

B. A. Bridges

M.R.C. Cell Mutation Unit, University of Sussex, Falmer, Brighton BN 1 9 QG

Summary

The last few years have given us new fundamental insights into the mechanisms of mutagenesis in bacteria, particularly *Escherichia coli.* The existence of a non-recombinational error-prone repair pathway is well established and has been implicated in mutagenesis by UV and ionizing radiations and by many chemicals (and photochemical reactions) that cause damage to DNA. It now seems increasingly likely that under normal conditions most lesions in DNA are not potentially mutagenic and only certain configurations or types of DNA damage are substrate for error-prone repair. Such configurations may, it now seems, be produced by certain base-analogue mutagens incorporated into DNA as well as by substances reacting chemically with it.

Error-prone repair has usually been studied in excision-deficient strains of bacteria but there is now evidence for the existence, in excision-proficient bacteria, of a hitherto unsuspected type of error-free excision-dependent post-replicative repair that may successfully compete with error-prone repair for potentially mutagenic lesions. The result is that at most loci, post-replicative repair is much less likely to lead to a mutation in wild type than in excision-deficient bacteria.

The precise mechanism of error-prone repair is as yet unknown but serious attempts are being made to identify the "error-prone enzyme". DNA polymerase III, terminal nucleotidyl transferase and reverse transcriptase have all been considered as possibilities.

The control of error-prone repair in bacteria is also intriguing. Error-prone activity appears to be "inducible" in a similar way to prophage induction. This is a convenient way of describing a complex phenomenon but it does not necessarily lead to increased understanding. That will only come when the steps of the "induction" process are elucidated; at the present time they are extremely controversial. An alternative to the hypothesis of "induction" of error-prone repair is presented as an Appendix.

"It is no paradox to say that in our most theoretical moods we may be nearest to our most practical applications".

A. N. Whitehead

This presentation is devoted largely to recent developments in our understanding of some of the ways in which base pair substitution mutations are produced in *Escherichia coli*. Other workers have recently published excellent reviews [16, 25, 42] and I therefore make no apology for the fact that much of the work to be discussed originates from my own laboratory, or for the fact that my emphases (and in some instances my interpretations) are different from theirs.

Most of the work to be discussed has involved ultraviolet (UV) or ionizing radiations although it is thought that many chemical mutagens will draw forth a similar response from the cell, probably with relatively minor variations. Before proceeding further, it is as well to state the basic model that grew up between 1965 and, say, 1975. The review of DOUDNEY [16], in particular, sets this out in some detail.

In this model damage to DNA is held to be dealt with by the bacterial cell in two main ways. Either the damage is removed by excision processes initiated by endonucleolytic or mutagen-induced breakage, usually before the region concerned is replicated, or the damaged region is replicated and the resulting discontinuities in newly-synthesized DNA (daughter strand gaps) are eliminated by means of post-replicational repair processes. The major post-replicational process seems to involve recombinational exchanges between the daughter strands [33, 34]. Both the excision-initiated and recombinational processes are largely error free, but error-prone repair (dependent upon the *lexA*$^+$ (= *exrA*$^+$) and *recA*$^+$ genes) can operate with a low probability either at excision gaps [6, 30, 31] or at daughter strand gaps [8, 26, 39]. Error-prone repair at the latter is much more common than at the former [22, 38].

It is the *lexA*$^+$-dependent mutagenic pathway with which I shall largely be concerned. Other modes of mutagenesis are of course recognized that are not *lexA*$^+$-dependent. They are believed to reflect miscoding at the time of replication either by a base analogue mechanism or because the damage to DNA is of a rather specific nature that subsequently alters the pairing specificity of a base [e.g. 27, 11, 19].

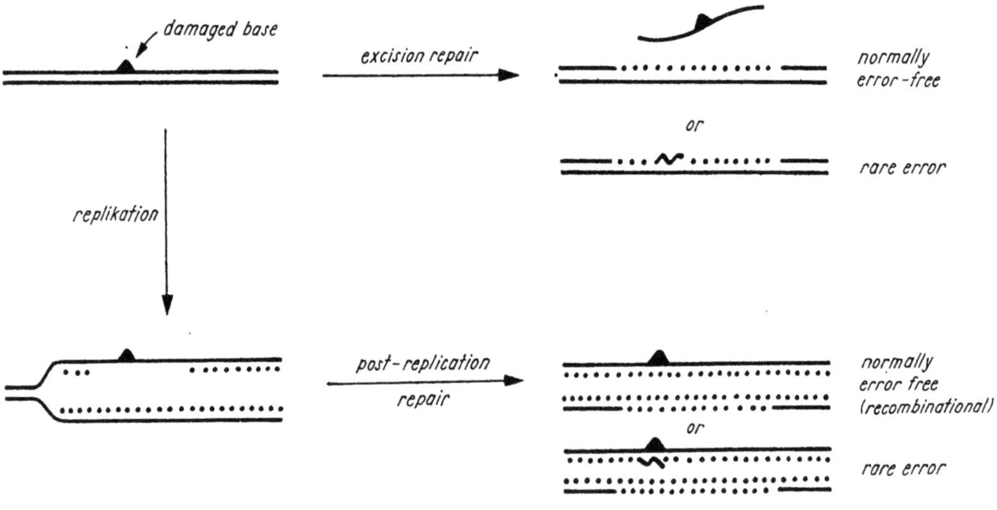

Fig. 1. DNA repair pathways in *E. coli*

Post-replication repair has usually been studied in bacteria deficient in excision repair (Uvr⁻) and it has been widely assumed that damage in excision-proficient bacteria which fails, for some reason or other, to be excised, is dealt with after DNA replication by means of similar processes to those operating in excision-deficient bacteria. Although the dependence of error-prone repair on the *recA*⁺ gene initially suggested that mutations might arise as errors in recombination [3, 25, 40] it is now clear that the *recA*⁺-dependence arises from the nature of the DNA damage and that the error-prone repair process does not require recombinational exchanges (see, for example, refs. 5 and 32).

This basic model is summarized in Fig. 1 and may still be regarded as a good working model. Recently, however, rather important refinements have come to light; these affect our understanding of (a) the nature of premutational lesions, (b) the pathways of error-free repair, and (c) the mechanism of error-prone repair.

Premutational lesions

The fact that at least 95% of both true and suppressor Trp⁺ mutants induced in WP2 *uvrA* by UV are photoreversible indicates that the initiating lesions are (or include) pyrimidine dimers. Mutations arise as a consequence of replication of DNA containing dimers but only with a rather low probability [8]. What is now emerging is that this low probability does not reflect a low but finite probability of error-prone repair at all daughter strand gaps, but rather a much higher probability at a small

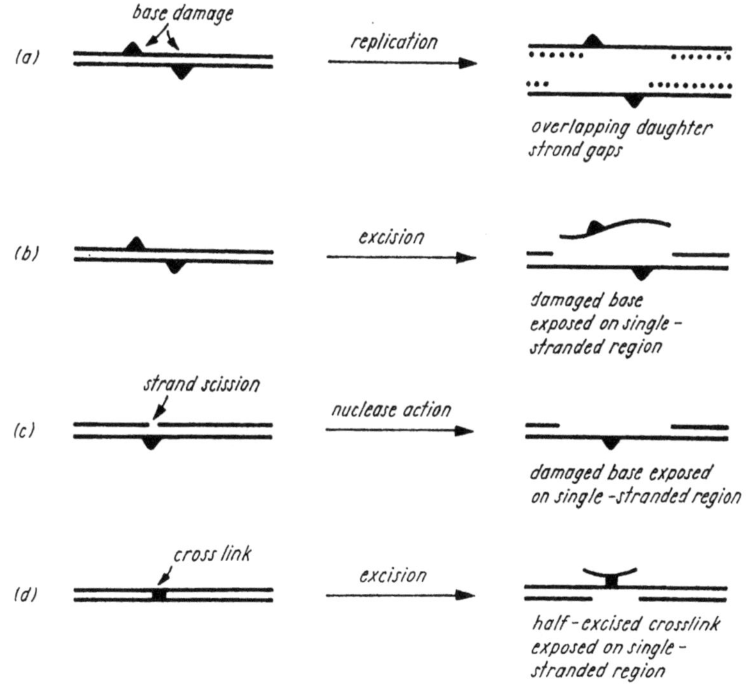

Fig. 2. Some DNA lesions that might have an enhanced or absolute requirement for error-prone repair.

specific fraction of such gaps [14]. Moreover, when error-prone repair does occur it precludes error-prone repair at the same site in subsequent replication cycles [14], i.e. it is a terminal event and does not leave behind any substrate for further error-prone repair. The possible nature of the premutational lesion in UV-irradiated excision-deficient bacteria was suggested by SEDGWICK [35]. He argued from the known requirement for two lesions [2, 15] that pyrimidine dimers close together on opposite strands of the DNA duplex would give rise, after replication of the region, to overlapping daughter strand gaps (ODSGs) (Fig. 2a). An ODSG would seem to be difficult to deal with by recombinational exchanges and might have an obligatory requirement for repair by the non-recombinational error-prone system. The ODSG model is also compatible with the observed terminal nature of error-prone repair since the pyrimidine dimers would normally tend to segregate and give rise to non-overlapping daughter strand gaps in different chromosomes at the next DNA replication [14].

Whereas most daughter strand gaps seem not to require error-prone repair, analogous configurations not reparable by normal polymerases might, if they occurred before DNA replication, be less easy to repair by recombination since the nearest homologous duplex might be spatially remote, for example in a separate nucleus. Such a configuration might exist where excision of radiation or chemically damaged base on one strand leads to the uncovering of a damaged base on the other strand (Fig. 2b), or where a single strand break induced by ionizing radiation is accompanied by base damage on the opposite strand, either produced by the same energy loss event or by two separate events (Fig. 2c). Such a possibility was put forward by BRESLER [1]. Excision of half of a DNA crosslink also results in a similar configuration (Fig. 2d).

In essence, our current view of premutational lesions (excluding those directly promoting errors at replication) is that they are exceptional and specific configurations involving radiation or chemically damaged sites that are difficult or impossible to repair by error-free processes. The corollary of this is that most types of DNA damage are normally dealt with by error-free processes and so do not lead to mutations. In exceptional circumstances they may be substrate for error-prone repair, for example in strains with certain plasmids or carrying a *tif* mutation (see below), or when error-free repair is inhibited (e.g. 7).

Hitherto, all mutagens acting via the lex^{+} dependent pathway have been thought to exert their effect by causing non-specific DNA damage which results in single stranded gaps in DNA some of which are substrate for error-prone repair. A few chemical mutagens (e.g. nitrosamines [27], captan hydrolysis product [11], sulphur half mustard [19]) appear to cause minor damage to bases which can alter their coding specificity and lead directly to replication errors without the involvement of error-prone repair. In some but not all instances such damaged bases are detectable by UV endonuclease. Base analogues such as 2-aminopurine and 5-bromouracil are also believed to cause replication errors and are not detectable by UV endonuclease.

Recently, however, we have obtained evidence: suggesting that there exists a class of base analogues that are, when incorporated into DNA, quite different in their properties. Benomyl (methyl-1-(butylcarbamoyl)-benzimidazole carbamate) is a widely used systemic fungicide which is, according to SEILER [37] incorporated into DNA after removal of the butylcarbamoyl group. We have recently shown that benomyl is mutagenic only through the *lex*$^+$-dependent error-prone repair pathway and that the premutagenic lesion is detectable by UV-endonuclease and is excised [24]. If SEILER is correct in his assertion that benzimidazole derivatives are base analogues then the incorporated base must be seen by the cell as a non-pairing purine with a large group attached to it. This is entirely consistent with its structure. The aromatic ring would not be expected to pair with other bases and should thus generate a daughter strand gap instead of a replication error. The carbamate group at the C-8 position, if it remained on the molecule when incorporated might appear similar to a large alkylation adduct and be recognized as such by UV endonuclease.

If this interpretation is correct, replication and generation of a daughter strand gap would not occur until one complete cycle after the incorporation of the base into DNA. This could, at least in principle, allow all incorporated bases to be excised and result in total absence of mutagenicity in Uvr$^+$ strains as has been observed both in *E. coli* [24] and in *Aspergillus nidulans* [21].

Error-free repair

It is known that potentially excisable DNA damage in an excision-proficient strain (Uvr$^+$) that for some reason fails to be excised before replication gives rise to daughter strand gaps and it has been generally assumed that these are treated in the same way as in an excision-deficient strain (Uvr$^-$) i.e., by error-free recombination repair and by error-prone repair. This was the impression created by much early work [26, 22, 38] which showed that at a given level of inactivation, the induced mutation frequency among the survivors of UV irradiation was very similar in excision-proficient and -deficient strains even though the doses received were markedly different. With hindsight we can now see that this is an erroneous impression created by the fact that all these workers were scoring tRNA suppressor mutants which, on broth-enriched plates, are quite atypical. On plates without broth enrichment excision-proficient strains are much less mutable at the same survival level than are excision-deficient strains [38]. We have recently pointed out that this latter observation is true for almost all other loci that have been examined [20], a fact which led us to the possibility that post-replication repair in wildtype strains is considerably (perhaps up to 200 times) less error-prone than in excision-deficient strains. I had been conscious for some time that the presence of UV endonuclease in Uvr$^+$ strains could make possible post-replication repair pathways denied to the Uvr$^-$ strains in which such processes are usually studied [36].

We therefore attempted to compare error-prone repair (as measured by mutation) with the number of daughter strand gaps generated in Uvr$^+$ and Uvr$^-$ bacteria following the same relatively low UV dose [20]. If one assumes that the rate of dimer excision in *E. coli* WP2 (Uvr$^+$) is exponential then from our measurements of initial excision rate we were able to calculate that there would be about 6 times more single daughter strand gaps and about 13 times more overlapping daughter strand gaps in WP2 *uvrA* than in WP2 (Uvr$^+$) given the same low UV dose. If excision were linear with time (and we could not distinguish this experimentally from an exponential process) then the figures would be 18 times and 27 times respectively. The mutation yields, however, differed by much larger factors than these, e.g. 175-fold for Trp$^+$ mutations after 3 Jm^{-2}

on plates without broth supplementation, and 25000-fold for Str^r mutations after 1 Jm^{-2}.

These figures are certainly an underestimate of the difference in error-proneness of daughter strand gap repair since many of the mutants in the Uvr$^+$ strains appear to arise as rare errors during repair of excision gaps before DNA replication. We therefore concluded, after detailed consideration of various other explanations, that overall daughter strand gap filling in Uvr$^+$ bacteria is substantially less error prone than in Uvr$^-$ bacteria. The most reasonable explanation for this is that there exists an additional error-free excision-dependent pathway of post-replication repair in Uvr$^+$ strains. At the present time we have no information as to the nature of this process, nor, since

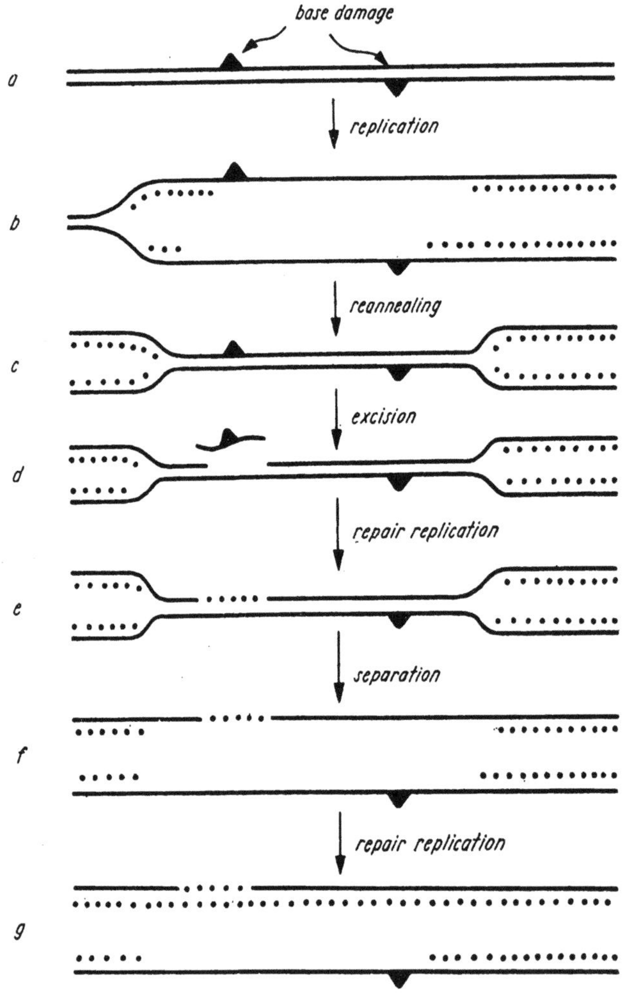

Fig. 3. Reannealing and post-replicative excision model for error-free excision-dependent post-replication repair. At configuration *e* the remaining base damage could be removed by a further excision step. Alternatively configuration *g* could be repaired by means of a recombinational exchange.

it operates only on a small fraction of DNA lesions, does it seem likely that a biochemical approach can be readily devised. We do, however, have a plausible model shown in Fig. 3.

This model supposes the reannealing of the parental strands after DNA replication, excision of a pyrimidine dimer (or, following chemical mutagenesis, of a damaged base) and the filling of the resultant gap using the original complementary strand as a template [20, 36]. Such a process need be involved only for the small specific proportion of daughter strand gaps which in Uvr⁻ bacteria are not susceptible to error-free recombination repair. Although the model could be applied to a single daughter strand gap by postulating the nicking and unwinding of a region of the sister chromosome, it works best with the ODSGs discussed above. In this case once the excision-dependent repair of one gap has occurred the gap on the sister chromosome would be amenable either to normal recombination repair or to a further excision event.

The probable existence of an excision-dependent mode of post-replication repair has thrown new light on one of the perennial problems of UV mutagenesis in *E. coli*, namely mutation frequency decline (MFD). MFD refers to the loss of mutations to prototrophy when bacteria are deprived of nutrient broth or casein hydrolysate after irradiation. It has been known for some time that this process is largely dependent on the excision repair system and is largely specific for nonsense suppressor mutations (see refs. 3 and 40). The mechanism has arosed considerable controversy (e.g. 12, 44) and none of the published models has ever seemed entirely satisfactory, all of them being based on differential rates of pre-replicative excision repair.

We have recently re-investigated the phenomenon and have shown that MFD can occur long after excision repair is complete. After considering possible alternatives we have concluded that MFD occurring at late times reflects an excision-dependent post-replicative process. Early MFD, known to be able to occur in the absence of DNA replication, may well reflect the rate of pre-replicative excision, but we would argue that even this may only be observed as MFD because it reduces the number of photoproducts left for subsequent replication and susceptibility to excision-dependent postreplication repair. The reannealing and post-replicative excision model provides an attractive explanation for late MFD since one needs only to postulate that in rich medium single-stranded *t*RNA genes opposite daughter strand gaps take up a self-annealing configuration that prevents reannealing with the other parental strand. The model "predicts" that, if postreplicational excision is inhibited at *t*RNA loci in rich media, then the error-proneness of post-replication repair will be elevated and tend to approach than in Uvr^- strains. This is observed experimentally when equal survival is used as the basis of comparison [38]. The fact that most of the early work on comparative UV mutagenesis in Uvr^+ and Uvr^- strains involved the scoring of suppressor mutants was, we believe, responsible for the failure to establish the low error-proneness of repair at most non *t*RNA loci in Uvr^+ strains. We have discussed this postreplication model for MFD in detail elsewhere [17]. Unfortunately is seems to share with all other ODSG models the disadvantage of being refractory to biochemical approaches.

Our evidence is not alone in pointing to a new error-free repair mechanism. MOUNT et al. [28] have obtained evidence for the appearance in UV-irradiated *E. coli* carrying *tsl* mutations (which are temperature-sensitive suppressors of *lexA*) of an error-free repair system capable of acting on bacteriophage λ. The *tsl* mutation has also been found to increase the UV resistance of *recA* strains [29]. The possibility therefore exists that *tsl* may promote reannealing and post-replicative excision in genetic backgrounds normally unfavourable for it (e.g. *recA lexA*).

Bacterial repair systems have usually been characterized by the isolation of mutations that block their operation. Excision-dependent postreplication repair is probably blocked in *recA* bacteria and might well be blocked in *lexA* bacteria although it is difficult to design experiments to prove either of these assertions. No mutants have so far been identified as specifically blocking the pathway although there are two which appear to interfere with it probably rather indirectly. The mutations *mfd* and $dnaB_{(1026)}$ have both been shown to cause a delay in excision of photoproducts after UV irradiation [4, 9, 18]. The delay does not alter survival after UV in either strain but UV-induced mutation rates are elevated, particularly at suppressor loci on media without nutrient broth. It has been suggested that these are properties expected of a mutant defective in reannealing and post-replicative excision and are hard to explain in any other way [9].

Error-prone repair

From the time it became clear that UV-induced mutations probably arose as errors in repair, two possibilities have been considered: either mutations arise as errors in recombination or as incorporation of wrong bases opposite non-coding damaged bases in DNA [8, 39]. Although the recombination error hypothesis was first favoured largely because of the requirement for the *recA*$^+$ gene, it is now clear that recombination is not necessary for error-prone repair (see 5). A further factor to be considered is the suggestion that error-prone repair is inducible [13, 32, 43], a suggestion that is still unproven and which I will consider further below.

There are three ways in which wrong bases could be inserted opposite non-coding lesions such as those shown in Fig. 2 [5].

1. There could be a new (possibly inducible) polymerase that has a lower template specificity than Pols I, II and III. Reverse transcriptase has been suggested for this role (Radman, personal communication).

2. There could be a (possibly inducible) terminal nucleotidyl transferase that can add bases to a single strand priming end without requiring any template [42].

3. One of the known polymerases could act in combination with a (possibly inducible) co-factor which could lower its template specificity. At the present time there is little evidence to distinguish between these hypotheses. Only with the third can some further deductions be made. One can, for example, rule out Pol I and Pol II because strains deficient in these polymerases are not deficient in UV mutagenesis. Polymerase III, in contrast, has been shown to be obligatory for error-prone repair [10]. Mutations induced by UV did not lose their photoreversibility in a *dnaE* temperature-sensitive strain (i.e., were not converted from pyrimidine dimers into irreversible mutations) at 43° even though normal replication and filling of daughter strand gaps still continued for some time. It was suggested that the polymerase was more temperature sensitive for its repair function than for its normal replication function. Certainly the isolated poly-

merase III (like that from many other *dnaE* strains) shows little or no temperature sensitivity for polymerizing activity when tested in vitro (A. FORTUNATO, unpublished data). Alternatively, CAIRNS (personal communication) has suggested that the stringent temperature sensitivity of Pol III for error-prone repair could be explained if free enzyme were required for this whereas normal replication involves bound enzyme which is thereby protected and able to continue replication for some time after being shifted to its restrictive temperature.

Although this result does not prove that Pol III is the enzyme needed to incorporate wrong bases, its obligatory involvement in error-prone repair does make this an attractive possibility.

There is currently, a generally held opinion that error-prone repair in *E. coli* is inducible, that is, the amount of error-prone enzyme(s) is increased in those bacteria that have been exposed to an inducing agent such as UV [32, 42]. Induction is supposed to be mechanistically similar to the induction of prophage λ, induction of WEIGLE reactivation, and mutagenesis of phage λ and the induction of filaments in *lon* strains, all processes dependent, like error-prone repair, upon the $recA^+$ and $lexA^+$ gene functions.

There is much that could be stated about the similarities and dissimilarities of these processes, but I propose only to discuss mutagenesis of bacteria. One of the major pieces of evidence in favour of inducible error-prone repair is the fact that bacteria carrying the temperature-sensitive *tif-1* mutation show greatly enhanced UV mutagenesis, particularly with low doses, if the bacteria are incubated at restrictive temperature for a period of time after UV (41). *Tif-1* bacteria spontaneously induce prophage λ, WEIGLE reactivation and mutagenesis of λ and filamentation at restrictive temperature. It was, therefore, logical to suppose that induction of the bacterial error-prone repair system also occurs at restrictive temperature. Unfortunately, this is not necessarily helpful since the mechanisms of these other processes are also unknown and the inducing signals in damaged DNA are obviously not identical for all the different effects. GREEN has pointed out (see e.g. in 5) that the phenomenon of "induction" in *tif-1* strains can be equally well explained in terms of an altered accessibility of DNA towards a constitutive error-prone enzyme. This hypothesis has not previously been published in detail and it is therefore given as an appendix to the present paper. Basically, it assumes that the cell needs some way of ensuring that single-strand DNA regions capable of being sealed by error-free processes (e.g. gaps in newly-replicated undamaged DNA and most daughter strand gaps opposite replicated regions containing damaged bases) are not accessible to error-prone enzymes. GREEN postulates that newly-replicated DNA is "immune" to error-prone repair for a certain period of time, after which those gaps not susceptible to error-free processes (repair replication or recombination) are still available and can act as substrate for error-prone repair. *Tif-1* bacteria are postulated to lift the "immunity" of newly-replicated DNA at restrictive temperature. Thus both daughter strand gaps and gaps formed normally during chromosomal replication would become available to error-prone enzymes, which would account for the high mutator activity of both irradiated and unirradiated *tif-1* strains at 42°.

The hypothesis also predicts that a substantial proportion of error-prone repair occurring before replication of a damaged region would not be delayed by "immunity" and would give the appearance of being constitutive. This is born out by observation; X-ray mutagenesis is not reduced by addition of chloramphenicol immediately after irradiation [23] and many mutations to Str^r (shown to arise at excision gaps) arise during liquid holding in buffer [6]. In both these cases one is obliged to conclude that "induction", in any meaningful sense, is ruled out.

In some recent studies with a *uvrA* non-filamenting *tif-1* strain derived from *E. coli* WP 2 and kindly given by Dr. Evelyn Witkin, I have obtained results which seem to me to be far more consistent with the "immunity" hypothesis than with the "induction" hypothesis. I have shown that 43° treatment after gamma radiation increases the number of induced mutants more than pre-treatment at 43° despite the fact that most of the known DNA damage is repaired very rapidly. One would have expected that any induction process begun after X-irradiation would take some time and would not be complete until after error-free processes had repaired most of the damage. Moreover, it was found that the full thermal enhancement could be obtained up to 2 h after a low dose of irradiation at which time all known DNA damage had been repaired and the bacteria were once again dividing exponentially.

The results are difficult to reconcile with the thermal induction of error-prone repair in *tif-1* bacteria. It seems preferable to postulate some long-lasting enhancement of error-prone potential by gamma irradiation which is only realised in heated *tif-1* bacteria when the DNA becomes accessible (loses its "immunity"?). The first of these processes could conceivably be thought of as induction of error-prone repair triggered by irradiation, but other, physiological, explanations can also be conceived. The *tif-1* mutation at 43° does not appear to exert its effect here by inducing a mutagenic activity but by allowing existing mutagenic activity, which may itself be increased by irradiation, to have access to DNA normally denied to it.

We have considered a similar explanation for the action of the mutation-enhancing plasmids, including those carried by the *Salmonella* tester strains TA 98 and TA 100. To test this possibility a derivative of *E. coli* WP 2 *uvrA* containing both the *tif-1* mutation and a mutation-enhancing plasmid pKM 101 has been constructed. Experiments have shown that the mutator effects of tif-1 and the plasmid are completely independent and more than additive, both for spontaneous and for UV-induced mutation (Doubleday, Bridges and Green, to be published). We therefore presume that *tif-1* and the pKM 101 plasmid act in entirely different ways to increase the yield of mutants.

I believe that the awakening of our interest in control of error-prone and error-free repair that stems from the work on Weigle reactivation and on *tif-1* strains is of great value. I am not, however, convinced that the first interpretation of inducibility of error-prone repair is the only one that should be considered. Only time will enable us to resolve the apparently conflicting ways of looking as these phenomena.

One point that emerges clearly from the *tif-1* and plasmid studies is that both these effects probably reflect, in their different ways, the production of mutations in undamaged DNA initiated by DNA damage which may not in itself normally be mutagenic. There is a moral here for those using plasmid-containing strains for short term testing. These strains are hypersensitive indicators of DNA damage. In so far as DNA damage correlates with carcinogenicity these strains may have value; but since most DNA damage is not even potentially mutagenic in normal bacteria, the use of plasmid-containing strains for screening for potential human *mutagens* (as distinct from *carcinogens*) may be quite unwarranted. This can, of course, only be determined when the mechanisms of repair in human cells are elucidated, but the possibility of false positives with plasmid strains as far as *genetic* risk is concerned must certainly exist.

To summarize the advances discussed in this paper, Fig. 4 shows diagrammatically the basic model of Fig. 1 revised in accordance with recent information.

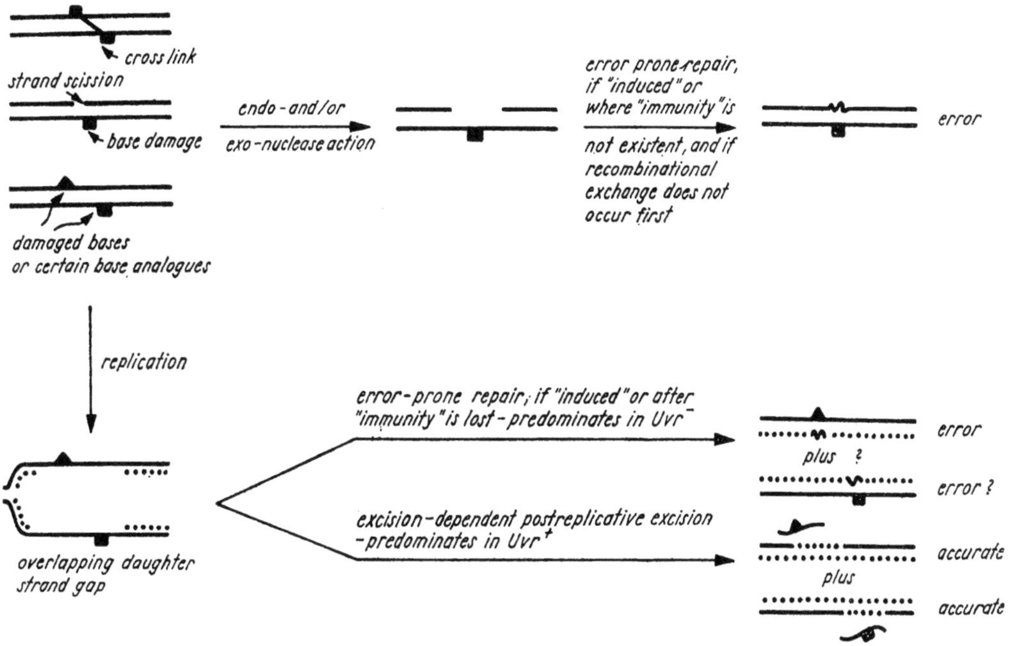

Fig. 4. A highly speculative revision of the model in Fig. 1 in which mutations result only from certain gap configurations not readily reparable by error-free excision repair or recombination repair. The latter two processes (to which most DNA lesions are susceptible) are omitted here.

Acknowledgements

Thanks are due to Michael GREEN for many helpful discussions and to Drs. Evelyn WITKIN and Charles DOUDNEY who kindly let me see their manuscripts before publication.

References

[1] BRESLER, S. E., Theory of misrepair mutagenesis, Mutation Res., **29** (1975) 467–474.

[2] BRIDGES, B. A., A note on the mechanism of UV mutagenesis in *Escherichia coli*, Mutation Res., **3** (1966) 273–279.

[3] BRIDGES, B. A., Mechanisms of radiation mutagenesis in cellular and subcellular systems, Ann. Rev. Nucl. Sci., **19** (1969) 139–178.

[4] BRIDGES, B. A., An ultraviolet resistant *ts* DNA mutant of *Escherichia coli* deficient in host cell reactivation ability for bacteriophage lambda and showing hypersensitivity towards induction of Weigle-reactivation, Mutation Res., **29** (1975) 489–492.

[5] BRIDGES, B. A., Genetic effects of UV on *Escherichia coli* – a model for prokaryotes, in Radiation Research, Biomedical, Chemical and Physical Perspectives, Academic Press, New York, 1975.

[6] BRIDGES, B. A., and R. P. MOTTERSHEAD, RecA$^+$-dependent mutagenesis occurring before DNA replication in UV- and γ-irradiated *Escherichia coli*, Mutation Res., **13** (1971) 1–8.

[7] BRIDGES, B. A., and R. P. MOTTERSHEAD, Gamma ray mutagenesis in a strain of *Escherichia coli* deficient in DNA polymerase I, Heredity, **29** (1972) 203–211.

[8] BRIDGES, B. A., R. E. DENNIS and R. J. MUNSON, Differential induction and repair of ultraviolet damage leading to true reversions and external suppressor mutations of an ochre codon in *Escherichia coli* B/r WP 2, Genetics, **57** (1967) 897–908.

[9] BRIDGES, B. A., A. R. LEHMANN, and R. P. MOTTERSHEAD, Error-prone DNA repair in *Escherichia coli*. IV. Excision repair and radiation-induced mutation in a *dnaB* strain, Biol. Zbl., **95** (1976) 393–403.

[10] BRIDGES, B. A., R. P. MOTTERSHEAD, and S. G. SEDGWICK, Mutagenic DNA repair in *Escherichia coli*. III. Requirement for a function of DNA polymerase III in ultraviolet light mutagenesis, Molec. gen. Genet., **144** (1976) 53–58.

[11] BRIDGES, B. A., R. P. MOTTERSHEAD, M. A. ROTHWELL and M. H. L. GREEN, Repair-deficient strains suitable for mutagenicity screening: tests with the fungicide captan, Chem.-Biol. Interactions, **5** (1972) 77–84.

[12] CLARKE, C. H., and R. F. HILL, Mutation frequency decline for streptomycin-resistant mutations induced by ultraviolet light in *Escherichia coli* B/r, Mutation Res., **14** (1972) 247–249.

[13] DEFAIS, M. P. FAUQUET, M. RADMAN and M. ERRERA, Ultraviolet reactivation and ultraviolet mutagenesis of λ in different genetic systems, Virology, **43** (1971) 495–503.

[14] DOUBLEDAY, O. P., B. A. BRIDGES and M. H. L. GREEN, Mutagenic DNA repair in *Escherichia coli*: II. Factors affecting loss of photoreversibility of UV-induced mutations, Molec. Gen. Genet., **140** (1975) 221–230.

[15] DOUDNEY, C. O., Ultraviolet light-induced mutation as an event in the physiology of the bacterial cell, in Repair of Genetic Radiation, ed. F. H. Sobels, Pergamon: Oxford (1963).

[16] DOUDNEY, C. O., Mutation in ultraviolet light-damaged micro-organisms, pp. 309–374. In: Photochemistry and Photobiology of Nucleic Acids, vol. 2., Academic Press: New York (1976).

[17] EYFJORD, J. E., M. H. L. GREEN, B. A. BRIDGES and W. J. MURIEL, Mutagenic DNA repair in *Escherichia coli*. V. The mechanism of mutation frequency decline, Mutation Res., in press (1976).

[18] GEORGE, D. L., and E. M. WITKIN, Slow excision repair in an *mfd* mutant of *Escherichia coli* B/r., Molec. gen. Genet., **133** (1974) 283–291.

[19] GILBERT, R. M., S. ROWLAND, C. L. DAVISON and B. PAPIRMEISTER, Involvement of separate pathways in the repair of mutational and lethal lesions induced by a monofunctional sulfur mustard, Mutation Res., **28** (1975) 257–275.

[20] GREEN, M. H. L., B. A. BRIDGES, J. E. EYFJORD and W. J. MURIEL, An error-free, excision-dependent mode of post-replication repair of DNA in *Escherichia coli*, in: Proceedings of International Colloquium Mecanismes d'alteration et de reparation du DNA: relations avec la mutagenese et la cancerogenese chimiques, Menton, 1976, in press.

[21] HASTIE, A. C., Benlate-induced instability of Aspergillus diploids, Nature, **226** (1970) 771.

[22] HILL, R. F., Ultraviolet-induced lethality and reversion to prototrophy in *Escherichia coli* strains with normal and reduced dark repair ability, Photochem. Photobiol. **4** (1965) 563–568.

[23] KADA, T., C. O. DOUDNEY and F. L. HAAS, Some biochemical factors in X-ray-induced mutation in bacteria, Genetics, **46** (1961) 683–702.

[24] KAPPAS, A., M. H. L. GREEN, B. A. BRIDGES, A. M. ROGERS and W. J. MURIEL, Benomyl – a novel type of base analogue mutagen? Mutation Res., in press (1976).

[25] KONDO, S., DNA repair and evolutionary considerations, in: Advances in Biophysics, Vol. 7, pp. 91–162 (ed. M. Kotani), University of Tokyo Press: Tokyo (1975).

[26] KONDO, S., H. ICHIKAWA, K. IWO and T. KATO, Base-change mutagenesis and prophage induction in strains of *Escherichia coli* with different DNA repair capacities, Genetics, **66** (1970) 187–217.

[27] LOVELESS, A., Possible relevance of 0−6 alkylation of deoxyguanosine to mutagenicity and carcinogenicity of nitrosamines and nitrosamides, Nature, **223** (1969) 206−207.
[28] MOUNT, D. W., C. K. KOSEL and A. WALKER, Inducible, error-free DNA repair in *tsl recA* mutants of *E. coli*, Molec. gen. Genet. **146** (1976) 37−41.
[29] MOUNT, D. W., A. C. WALKER and C. KOSEL, Effect of *tsl* mutations in decreasing radiation sensitivity of a *recA*$^-$ strain of *E. coli* K-12, J. Bact., **121** (1975) 1203−1207.
[30] NISHIOKA, H., and C. O. DOUDNEY, Different modes of loss of photoreversibility of mutation and lethal damage in ultraviolet light-resistant and sensitive bacteria, Mutation Res., **8** (1969) 215−228.
[31] NISHIOKA, H., and C. O. DOUDNEY, Different modes of loss of photoreversibility of ultraviolet light-induced true and suppressor mutations to tryptophan independence in an auxotrophic strain of *Escherichia coli*, Mutation Res., **9** (1970) 349−358.
[32] RADMAN, M., SOS repair hypothesis: phenomenology of an inducible DNA repair which is accompanied by mutagenesis, in: Molecular Mechanisms of Repair of DNA (ed. P. C. Hanawalt and R. B. Setlow) pp. 355−367, Plenum: New York (1975).
[33] RUPP, W. D., and P. HOWARD-FLANDERS, Discontinuities in the DNA synthesized in an excision-defective strain of *Escherichia coli* following ultraviolet irradiation, J. Mol. Biol. **31** (1968) 291−304.
[34] RUPP, W. D., C. E. WILDE, D. L. RENO, and P. HOWARD-FLANDERS, Exchanges between DNA strands in ultraviolet irradiated *Escherichia coli*, J. Mol. Biol., **61** (1971) 25−44.
[35] SEDGWICK, S. G., Genetic and kinetic evidence for different types of post-replication repair in *Escherichia coli* B, J. Bact., **123** (1975) 154−161.
[36] SEDGWICK, S. G., and B. A. BRIDGES, Requirement for either DNA polymerase I or DNA polymerase III in post-replication repair in excision-proficient *Escherichia coli*, Nature **249** (1974) 348−349.
[37] SEILER, J. P., Toxicology and genetic effects of benzimidazole compounds, Mutation Res., **32** (1975) 161−168.
[38] WITKIN, E. M., Mutation and the repair of radiation damage in bacteria, Radiation Res. Supplement **6** (1966) 30−53.
[39] WITKIN, E. M., Mutation-proof and mutation-prone modes of survival in derivatives of *Escherichia coli* B differing in sensitivity to ultraviolet light, Brookhaven Symp. Biol., **20** (1967) 17−55.
[40] WITKIN, E. M., Ultraviolet-induced mutation and DNA repair, Ann. Rev. Genet., **3** (1969) 525−552.
[41] WITKIN, E. M., Thermal enhancement of ultraviolet mutability in a *tif-1 uvrA* derivative of *E. coli* B/r: evidence that ultraviolet mutagenesis depends upon an inducible function, Proc. Nat. Acad. Sci. USA, **71** (1974) 1930−1934.
[42] WITKIN, E. M., Ultraviolet mutagenesis and inducible DNA repair in *Escherichia coli*, Bact. Rev., in press (1976).
[43] WITKIN, E. M., and D. L. GEORGE, Ultraviolet mutagenesis in *polA* and *uvrA polA* derivatives of *E. coli* B/r: evidence for an inducible error-prone repair system, Genetics Suppl., **13** (1973) 91−108.
[44] WITKIN, E. M., and I. E. WERMUNDSEN, Do ultraviolet-induced mutations to streptomycin resistance exhibit susceptibility to mutation frequency decline?, Mutation Res., **19** (1973) 261−264.

Appendix

On the Possible Immunity of Newly-Synthesized DNA to Error-Prone Repair

M. H. L. GREEN

MRC Cell Mutation Unit, University of Sussex

One of the most useful contributions to our understanding of UV mutagenesis, has been the suggestion by RADMAN [7] that the error-prone repair process involved in mutation by agents such as UV (which he terms SOS-repair) is itself induced by DNA damage. The principal evidence for this hypothesis comes from (a) the phenomenon of UV-reactivation, which shows the exact characteristics of an inducible error-prone repair system acting on infecting phage DNA, (b) the *tif-1* mutant, which behaves as if constitutive for error-prone repair at 42 °C, and (c) the dose-squared induction of many types of UV-induced mutation, which can be interpreted as indicating a need both to generate lesions and induce error-prone repair in order to obtain mutation [9].

Other evidence, however, is less easily explained by the hypothesis. Firstly, it would be predicted from UV reactivation that conditions favouring high bacterial mutagenesis would increase survival. In fact, growth conditions greatly influencing bacterial mutagenesis generally have little effect on survival and in certain cases (e.g. K-12 *uvr* strains [5], *tif uvr* strains (J. GEORGE, pers. comm.) there is a much lower survival under conditions increasing mutagenesis. Moreover, in our hands split dose experiments [4] and experiments with irradiated plasmids [1] have not provided evidence of an inducible process affecting *bacterial* survival or mutagenesis.

Secondly the results with gamma-irradiated *tif-1* bacteria described in the preceding paper suggest that the classical interpretation of its effects with UV may not apply. Thirdly, the currently more popular interpretation of dose-squared mutation induction is that a specific fraction of lesions interact with each other to prevent their error-free repair [2, 8]. Inducible error-prone repair is not an essential component of this interpretation.

Finally, although UV-induced mutation occurs with a considerable lag and a requirement for protein synthesis (suggesting a need for enzyme induction), γ-induced mutation shows little lag or requirement for protein synthesis [6].

These considerations have led me to suggest an alternative hypothesis, that the enzymes required for error-prone repair are constitutive, but that they are unable to act on newly-synthesized DNA (i.e. it is IMMUNE to error-prone repair).

Obviously this model has advantages in explaining the discrepancy in kinetics between UV mutagenesis (in newly-synthesized DNA) and γ mutagenesis (in pre-existing DNA). It also accounts for the discrepancies between increased mutagenesis and increased survival after UV. Recombination repair can occur at a time when error-prone repair is prevented [3] so that conditions favouring recombination repair will enhance survival and reduce mutation. Finally the model reconciles the evidence that error-prone repair after UV is delayed [3] with the lack of evidence for inducible repair in our split dose experiments [4].

The model can be also extended to encompass *tif-1* as a strain in which the immunity of newly-synthesized DNA is lost at 42°, so that all gaps, including OKAZAKI gaps, are a substrate for error-prone repair. Thus both the spontaneous and induced mutator effects in *tif-1* bacteria can be accounted for. Even UV-reactivation can be explained if phage DNA, because of its smaller size, never escapes immunity unless the cell is irradiated.

I do not consider the hypothesis of immunity in any way proved, and I merely suggest it as an alternative to inducible error-prone repair. The mechanism of error-prone repair will be ultimately determined by characterizing the actual enzymes involved and neither inducibility nor immunity may turn out to correspond particularly closely to the actual situation. There are obviously many possibilities and too narrow a preconceived idea of the nature of error-prone repair may hinder the search for the detailed mechanism.

References

[1] BRIDGES, B. A., M. A. ROTHWELL and M. H. L. GREEN, Repair processes and dose response curves in ultraviolet mutagenesis of bacteria, Anais da Academia Brasiliera de Ciencias **45** (suppl.) (1973) 203–209.

[2] DOUBLEDAY, O. P., B. A. BRIDGES and M. H. L. GREEN, Mutagenic DNA repair in *Escherichia coli*. II. Factors affecting loss of photoreversibility of UV-induced mutations, Molec. gen. Genet., **140** (1975) 221–230.

[3] EYFJÖRD, J. E., M. H. L. GREEN and B. A. BRIDGES, Mutagenic DNA repair in *Escherichia coli*: Conditions for error-free filling of daughter strand gaps, J. Gen. Microbiol. **91** (1975) 369–375.

[4] GREEN, M. H. L., B. A. BRIDGES, J. E. EYFJÖRD and W. J. MURIEL, An error-free excision-dependent mode of post-replication repair in *Escherichia coli*, Prodeedings Colloque International CNRS "Mécanismes d'altération et de réparation du DNA: relations avec la mutagénèse et la cancérogénèse chimiques" (1976).

[5] GREEN, M. H. L., M. A. ROTHWELL and B. A. BRIDGES, Mutation to prototrophy in *Escherichia coli* K-12: effect of broth on UV-induced mutations in strain AB 1157 and four excision-deficient mutants, Mutation Res., **16** (1972) 225–234.

[6] KADA, T., C. O. DOUDNEY and F. L. HAAS, Some biochemical factors in X-ray induced mutation in bacteria, Genetics, **46** (1961) 683–702.

[7] RADMAN, M., SOS repair hypothesis: phenomenology of an inducible DNA repair which is accompanied by mutagenesis, in: Molecular Mechanisms of Repair of DNA (ed. P. C. Hanawalt and R. B. Setlow) pp. 355–367, Plenum: New York (1975).

[8] SEDGWICK, S. G., Genetic and kinetic evidence for different types of post-replication repair in *Escherichia coli* B, J. Bacteriol. **123** (1975) 154–161.

[9] WITKIN, E. M., Elevated mutability of *polA* and *uvrA polA* derivatives of *Escherichia coli* B/r at sublethal doses of ultraviolet light: Evidence for an inducible error-prone repair system ("SOS" repair) and its anomalous expression in these strains, Genetics, **79** (1975) 199–213.

Perspective on the Mutagenicity of Chemical Carcinogens

Frederick J. DE SERRES

National Institute of Environmental Health Sciences
Research Triangle Park, NC 27709

The existance of a correlation between carcinogenic and mutagenic activities is not only a problem of great theoretical interest, but one of great practical importance, if indeed, we have discovered a simple means to detect and eliminate potential mutagens and carcinogens in our environment. In addition, it is also possible that the production of high levels of genetic damage in high-risk, worker populations may have predictive value for carcinogenic effects which may not develop until 15 or 20 years later.

Basically what I hope to accomplish in this paper is:

(1) to review what we have learned about the mutagenicity of chemical carcinogens,

(2) to present new data which shows a quantitative correlation between carcinogenic and mutagenic activity as well as their general significance, and

(3) to discuss the appropriate use of this new technology for reduction of genetic disease and cancer in our society.

Correlation between carcinogenic and mutagenic activity

The data presented in the McCANN and AMES paper (1976) on tests with *Salmonella* clearly show that 90% (157/175 chemical carcinogens tested) are mutagens, whereas only 13% (14/108 chemicals believed to be noncarcinogenic) gave a similar positive response (Tab. 1). It is not generally recognized that these data represent studies done

Table 1

Mutagenicity of chemical carcinogens and noncarcinogens[1]

	Total Tested	Total Positive	Percentage
Carcinogens	175	157	89.7
Noncarcinogens	62[2]	14	
	46[3]	0	
	108	14	13.0

[1] adapted from McCANN et al. (1975), McCANN and AMES (1976)

[2] includes chemicals closely related and even isomeric to known carcinogens

[3] common biochemicals

not only in AMES' laboratory but in many other laboratories and that this tabulation includes data on at least 60 chemicals from a Japanese study which was started independently in 1972 (McCANN et al., 1975). During this present EEMS meeting we have heard of another completely independent study started at the Imperial Chemical Industries, Limited at Alderley Park, England, which both confirms and extends these earlier observations. All of these studies have produced two classes of anomalous data:

(1) False negatives and (2) False positives

False Negatives — In the studies with *Salmonella* by AMES group (Table 2) these are chemical carcinogens which did not give a positive response with *Salmonella*. There is a general feeling that the utilization of additional test systems might be particularly useful to determine whether other approaches might give a positive response. In examining the data that is available from other assay systems it is disappointing to see how few data have been collected by other investigators during the past year.

Table 2

Non-mutagenic carcinogens[1]

Code Number	Name
A 41	O-toluidine
A 43	Auramine
B 2	Carbon tetrachloride
B 7	2-chlorobutadiene (chloroprene)
B 24	Dieldrin
D 19	Thiourea
D 21	Ethyl carbamate (urethane)
D 27	Thioacetamide
D 28	Acetamide
F 7	Ethionine
F 15	Safrole
G 21	Cycasin
J 1	Phenobarbital, sodium
J 3	4-amino-2,3-dimethyl-1-phenyl-3-pyrazolin-5-one (4-aminoantipyrene)
J 5	3-amino-1H-1,2,4-triazole (amitrole)
K 5	1,2-dimethylhydrazine
K 7	Natulan (procarbazine)

[1] adapted from McCANN and AMES (1976)

ROSENKRANZ's assay for DNA repair (Table 3) gives positive results with o-toluidine, auramine, safrole, cycasin, 4-amino antipyrene, 1,2, dimethylhydrazine and natulan. KADA with his rec assay also gets positive results with thioacetamide and ethionine. Ethionine is also positive in *E. coli* strain WP-2 in SUGIMURA's laboratory and both safrole and ethionine have been found to give positive responses by Ann MITCHELL at Stanford Research Institute in tests for unscheduled DNA synthesis. Using in vitro transformation on golden syrian hamster embryo cells in culture Roman PIENTA at the Frederick Cancer Research Center has found that thiourea, thioacetamide, acetamide, ethionine and safrole all show in vitro transformation.

Table 3

Response of chemical carcinogens inactive in *Salmonella* with other assay systems

Compound[7]	DNA Repair[1] Pol$^+$/pol$^-$	DNA Repair[2] Rec Assay	Mitotic Recombination in yeast[3]	Mutation Induction *E. coli* WP-2[4]	Unscheduled DNA Synthesis[5]	In vitro Transformation[6]	*Salmonella*[7] hist$^-$→hist$^+$
A 41	+						–
A 43	+					–	–
B 2	?						–
B 7	?						–
B 24	?						–
D 19	–		–			+	–
D 21	–	–	–			–	–
D 27	–	+	+			+	–
D 28	–		–			+	–
F 7	–	+	–	+	+ (M.A.)[9]	+	–
F 15	+		+/–		+ (M.A.) –	+	–
G 21	+	+ (HMA)[8]					–
J 1	–						–
J 3	+?						–
J 5	–		–				–
K 5	+		–			+	–
K 7	+		+		+ (M.A.)	–	–

[1] Unpublished data, H. ROSENKRANZ
[2] Unpublished data, T. KADA
[3] Unpublished data, D. BRUSICK
[4] Unpublished data, T. SUGIMURA
[5] Unpublished data, A. MITCHELL
[6] Unpublished data, R. PIENTA
[7] J. McCANN and B. AMES (1976)
[8] Host-mediated assay
[9] Metabolic activation

These data are limited, but the message is clear that a battery of tests shows a higher correlation than the *Salmonella* test used in isolation. *False Positives* — In the studies with *Salmonella* by AMES' group (Table 4) these are chemicals which are believed to be noncarcinogens which gave a positive test with *Salmonella*. Obviously we have two alternatives once the problem of impurities in the samples has been resolved.

Table 4

Noncarcinogenic mutagens[1]

Code Number	Name	Revertants/mMole
A 11	4-acetylaminofluorene	0.3
A 22	α-naphthylamine	0,42
B 18	2-methoxy-6-chloro-9-[3-(2-chloroethyl) aminopropyl-amino]acridine · 2HCl(ICR-191)	511
B 22	Folpet	64
C 8	Benzo(e)pyrene	0.60
C 12	Dibenz(a, h)anthracene-5,6-oxide	5.3
D 14	2,3-epoxy-1-propanol (glycidol)	0.58
D 16	1,2-epoxybutane	0.01
E 18[2]	5-nitro-2-furamidoxime	5.3
E 19[2]	1-[(5-nitrofurfurylidene)-amino]-hydantoin	230
E 20[2]	5-nitro-2-furoic acid	0.26
K 11[2]	Sodium Azide	150
K 13	Sodium nitrite	0.01
L 9	N-hydroxy-4-aminoazobenzene	0.35

[1] adapted from McCANN and AMES (1976)
[2] Promoters

(1) True false positives which indicated problems with this overall approach or problems with the assay system or

(2) That these chemicals are truly carcinogens.

Examination of the data from other assay systems (Table 5) again is disappointingly meager, but they show that folpet, for example gives a positive response in the rec assay test and in the unscheduled DNA synthesis assay. Glycidol is also positive in the repair test, in yeast, and in the in vitro transformation test. Both sodium azide and sodium nitrite are known to be mutagenic in a large number of other assay systems. This confirmation of mutagenic activity suggests that the first alternative is clearly not correct in some cases. Some of these chemicals are genetically active in a number of assay systems. Perhaps if they are only weak carcinogens more extensive animal data need to be collected to evaluate their carcinogenic potential more thoroughly.

General problem of false positives

The general problem area of false positives has lead several research groups, namely Matthew MESELSON's at Harvard University as well as Bruce AMES' group at the University of California, Berkeley, to look into the possibility of a quantitative correlation between carcinogenic and mutagenic activity. During the past few months there has

Table 5

Response of noncarcinogenic chemicals active in salmonella with other assay systems

Compound[7]	DNA Repair[2] Pol$^+$/pol$^-$	DNA Repair[2] Rec Assay	Mitotic Recombination in yeast[3]	Mutation Induction *E. coli* WP-2[4]	Unscheduled DNA Synthesis[5]	In vitro Transformation[6]	*Salmonella*[7] hist$^-$ → hist$^+$
A 11	−		+				+
A 22	−		−		−	−	+
B 18							+
B 22	−	+	+		+ (M.A.)[8]		+
C 8	−		−			−	+
C 12							+
D 14	+		+		−	+	+
D 16	−		+				+
E 18		+		+/−			+
E 19	+						+
E 20		−		+			+
K 11	+	+					+
K 13	−	+		+			+
L 9							+

[1] Unpublished data, H. Rosenkranz
[2] Unpublished data, T. Kada
[3] Unpublished data, D. Brusick
[4] Unpublished data, T. Sugimura
[5] Unpublished data, A. Mitchell
[6] Unpublished data, R. Pienta
[7] J. McCann and B. Ames (1976)
[8] Metabolic activation

been extensive debate over these attempts to demonstrate a quantitive correlation between carcinogenic and mutagenic activity in the United States. At the Cold Spring Harbor Symposium on "The Origins of Human Cancer" in September 1976, data which was presented by RUSSEL and MESELSON (in press) on 10 carcinogens showed a precise correlation between carcinogenic and mutagenic activity. Nitrosamines don't fit well and were omitted from their data plot.

AMES group (McCANN, SAWYER and AMES, in press) has developed a similar plot with about 30 carcinogens which also shows that there is a good correlation with many compounds. Chlorinated hydrocarbons show no correlation since they are not mutagenic in *Salmonella.* Both safrole and urethane are also negative in *Salmonella* in Ames' hands and since they were positive in the ICI study it would be interesting to see how their position would be modified on the Ames data plot.

Nitrofurans (FANFT) also give a poor correlation — too many revertant colonies were found in the *Salmonella* test. These chemicals are most probably superactivated by *Salmonella* since it has three different nitroreductases. Herbert ROSENKRANZ (unpublished data) has shown that in nitroreductase deficient bacteria other nitrofurans actually give fewer revertants. Tests on ROSENKRANZ's strains would probably shift the FANFT data point to the left on the AMES data plot. Several attempts have also been made by this group to plot human data obtained from cigarette smoke condensate, data from roofers who work with melted tar as well as data on aflatoxin B_1 induced carcinogenesis. The correlation is good if we use both animal data on carcinogenicity as well as what little human test data are available. Nitrosamines were also eliminated on this plot because they show second order reactions rather than a linear reaction with the liver microsomes. Thus with nitrosamines you get a higher frequency of revertants than one gets in tests on chemicals which show linear dose response curves. With these restrictions and limitations both the RUSSELL and MESELSON plot and the McCANN, SAWYER and AMES plot give the appearance of some sort of quantitative correlation between carcinogenic and mutagenic activity.

Illusion or reality

Obviously the data showing a quantitative correlation between carcinogenic and mutagenic activity are somewhat biased; particular chemical classes have been omitted for a variety of reasons, and both the data on carcinogenicity and mutagenicity of a given chemical have been stringently selected. In theory, estimation of mutagenic potency for a given chemical is extremely difficult in *Salmonella* since the number of revertants is markedly strain-dependent. In testing the same concentration of a given chemical in strain G 46 and its various derivatives, including TA 100, one can obtain a marked apparent increase in potency over several orders of magnitude. In addition it is well known that carcinogenicity varies markedly in different mammalian species. Taken at face value, however, some chemicals do show a high correlation between carcinogenic and mutagenic activity. Obviously such a correlation does not exist for all chemicals and other scientists visualize a more complex correlation which shows a marked dependence on chemical class. Also the correlation will probably depend on whether the carcinogen is direct acting or requires metabolic activation. Obviously much more work needs to be done to get more data as a function of chemical class both with regard to mutagenicity data as well as carcinogenicity data. I believe that we are seriously hampered in evaluating these attempts to evaluate such a quantitative correlation by a totally inadequate data base.

As scientists we are concerned that present studies have given the illusion that (1) because there is a high correlation between carcinogenic and mutagenic activity and (2) that there is a correlation between mutagenic and carcinogenic potency. Substantial publicity has been given to these data in the popular press in many countries and there is the impression that the number of colonies observed in the *Salmonella* assay indicates not only mutagenic potency in man but carcinogenic potency as well!

What do we actually know about the carcinogenicity of chemical mutagens? Again we have too little data to give anything more than an estimate. Estimates vary from a simplistic 90% to a more cautious 10—15%. In fact, we really do not know. There is a strong possibility that mutagens are carcinogens but we do not know the probability. Nowhere does this leave us with regard to the effective utilization of present technology and the significance of our present data base?

Crisis stage

Two recent events clearly indicate that research in environmental mutagenesis has reached a crisis that will require aggressive action on the parts of members of all Environmental Mutagen Societies to bring these matters into proper focus and perspective. The first is the banning of hair dye components by the Italian government on June 18, 1976 (FALCO, 1976). The second is an equally aggressive attempt to ban flame retardants (BLUM and AMES, in press) as potentially potent carcinogens on the basis of positive tests for mutagenicity in *Salmonella*.

In the same way that we cannot answer the question about carcinogenic risk in workers with high frequencies of chromosome aberrations in lymphocyte culture, we can not as yet evaluate carcinogenic risk from mutagenicity tests on microorganisms. That is not to say that we may not be able to do it in the future, but we cannot do it with confidence today! Our experience is too limited and our data base much too inadequate. We must not underestimate the impact of erroneous prediction on our long term aims and objectives in this area of research. There is no doubt that too many such situations of what may turn out to be false alarms based on inadequate scientific data will eventually lead to a loss of credibility that will effectively destroy much of what we have worked so hard to create. But we still have not evolved an effective mechanism to cope with our over zealous colleagues who claim that *Salmonella* test can be used now to detect human carcinogens and mutagens in our environment. These new discoveries can have an enormous impact on our approach to human health problems if they are properly used.

We now have an effective mechanism to screen large number of untested environmental chemicals and to indentify those with mutagenic and carcinogenic potential. I stress the work potential, since I firmly believe that in general the *Salmonella* test is only a qualitative indicator of mutagenic activity. Nevertheless the use of such tests will provide an unprecedented opportunity to prioritize further testing of those compounds which give positive results in mass screening programs. Follow-up studies in higher organisms are required to provide the type of data required for informed risk analysis. Many of us believe that we cannot evaluate risk for either the mutagenic or carcinogenic hazard from the type of data produced in the short-term tests for mutagenicity at this time. Additional testing is required which gives more adequate data on potency as well as the spectrum of genetic alterations produced. Even with these data our experience in such extrapolation is so limited that rules have not as yet been developed either to estimate risk for cancer from genetic damage produced in somatic cells or to estimate risk for future generations from damage produced in germ cells. Obviously the rules must be different for an agent that produced exclusively nondisjunction than from an agent which produces point mutations by base-pair substitution! With time, guidelines for evaluating risk will be developed. The potential is there and unlike other areas of toxicology we have the means to obtain precise quantitative and qualitative data for all types of genetic damage. We can also predict the probability of successful transmission of particular types of genetic damage as well as the consequences of successful transmission in terms of genetic disease. In fact I believe that we even have the potential to estimate the differences in the frequencies and types of damage that might be expected from exposure to a mutagenic agent in our genetically heterogeneous human population. All people are not identical with regard to their ability to activate and detoxify toxic chemicals as is well exemplified by the marked differences which have been observed in the mutagenic and carcinogenic response to such industrial chemicals as vinyl chloride.

Future Course — My final comments are directed towards the future course of research work in environmental mutagenesis and the various projects which in my judgment should be given the highest possible priority.

(1) Every effort must be made to coördinate data collection on selected lists of chemicals to broaden our experience with the eccentricities of different chemical classes. We simply do not have an adequate data base or a sufficiently broad background of experience to tackle the enormous task of mass screening programs. Too often chemicals have been deliberately selected because it was known or suspected that they would give a positive test. Too many others which give troublesome test data have essentially been ignored.

(2) New assay systems must be developed to build redundancy in our mass screening programs. If we encounter chemicals which produce a highly specific effect we should be able to confirm it by a similar assay in some other indicator organism. This is a particularly important point with regard to developing a truly comprehensive screen for mutagenic activity because of the variety of effects that can be produced.

(3) We must also give a high priority to research on method development for population monitoring. I believe our most pressing problem in this area is future exploration of methodology for identifying susceptible individuals in high-risk worker populations as well as other high-risk populations. It is unrealistic to think in terms of banning the production of all active chemicals. We certainly can reduce levels of exposure with better industrial hygiene, but we can also hope to develop genetic tests to sort out the sensitive individuals with inadequate capabilities for either enzymatic detoxification or repair.

(4) More rapid means must be developed for communication between geographically isolated groups of research workers in the rapidly developing field. This includes rapid means for communication of new technical developments as well as new test data. We have too much work to do to waste inadequate resources repeating needlessly work already done adequately by other research groups.

(5) These additional means must also be developed for training new personnel as well as to provide rapid training of established workers as new approaches and techniques are developed. We are not dealing with problems that are unique to any particular country. These are world-wide problems where the developed countries must take the initiative for the rest of the world.

Finally it is clear that we have created a new field in toxicology and with this comes the responsibility to establish the general principles and guidelines by which to collect and utilize testing data.

I was pleased to see the great strides that had been taken along these lines at this meeting as well as the future plans of the European Environmental Mutagen Society. The enthusiasm, productivity and sense of purpose shown by members of this Society provides an excellent example for other scientists working in this area of research throughout the world.

References

[1] BLUM, A. and B. N. AMES (in press), Flame Retardants Additives as Possible Cancer Hazards. Science.

[2] FALCO, D. (1976), Prohibition of use of Hair Dyes with Coloring Substances Endowed with Mutagenic Capacity, Decree of the Minister of Health on June 18, 1976, Official Gazette of the Italian Republic No. 166 (June 25, 1976) p. 5025.

[3] MCCANN, J. and B. N. AMES, Detection of Carcinogens as Mutagens in the *Salmonella*/Microsome Test: Assays of 300 Chemicals; Discussion. Proc. Natl. Acad. Sci. USA **73** (1976) 950–954.

[4] MCCANN, J., E. CHOI, E. YAMASAKI and B. N. AMES, Detection of Carcinogens as Mutagens in the *Salmonella*/Microsome Test: Assays of 300 Chemicals, Proc. Natl. Acad. Sci. USA **72** (1975) 5135–5139.

[5] MCCANN, J., C. SAWYER and B. N. AMES (in press), The *Salmonella*/Microsome Test: Predictive Value for Animal Carcinogenicity, Cold Spring Harbor Symposium, Quant. Biol.

[6] RUSSELL, K. and M. MESELSON (in press), Quantitative Measures of Carcinogenic and Mutagenic Potency, Cold Spring Harbor Symposium, Quant. Biol.

Symposium I

Modifying Effects in Chemical Mutagenesis

Chairman's Introduction

N. P. DUBININ

Institute of General Genetics, USSR Academy of Science, Moscow

Modification of the mutational process utilizes primarily the multistage nature of the mutation appearance.. The stages are: transport of chemical mutagens into the cell, appearance of primary lesions in DNA, repair and replication of lesions, fixation of mutations.

Modification of transport. The effect of chemical mutagen depends on its concentration in the cell. Permeability of cell and nuclear membranes through which the mutagen molecules pass may be differential. Mutagen molecules may be intercepted by cytoplasmic proteins.

Enzymatic modification. Enzymes are capable to alter the metabolized compounds, converting non-mutagenic substances into mutagens, amplifying or weakening the mutagen effects (AMES et al., 1973; MALLING, 1971, 1974; GLETTEN et al., 1975).

Modification of mutagen reactions with DNA. An example is offered by hydroxylamine whose action type depends on concentration, pH, temperature, osmotic pressure, O_2 concentration, etc. (FREESE and FREESE, 1965; SINGER and FRAENKEL-CONRAT, 1969; BUDOWSKY and SVERDLOV, 1972).

Hydroxylamine was shown to react with free cytosine forming hydro- and aminocytidines whose incorporation into DNA results in appearance of mutations (CHU et al., 1974; SALGANIK et al., 1973).

Mutagenic and lethal effects of nitrosoguanidine are pH-dependent (KOLB and KAUDEWITZ, 1970; CERDA-OLMEDO and HANAWALT, 1968).

Modification of replication and repair systems. Replication and repair were shown to have a considerable impact in chemical and radiation mutagenesis. Different types of replication and repair are modifyed to different extent.

Two types of pre-replicational repair are known: 1) several nucleotides per lesion are incorporated during the repair synthesis of DNA. The process is controlled by *uvr* and *pol A* genes; 2) about 1000 nucleotides are incorporated per lesion, *rec A* and *rec B* being responsible (COOPER, HANAWALT, 1972).

Post-replication repair seals the gap that appears on the opposite strand of the lesion during its replication (RUPP, HOWARD-FLANDERS, 1968; LEHMAN, 1972; TROSKO et al., 1973). There is also an inducible system of post-replicational repair (WITKIN, GEORGE 1973). There are systems that remove crosslinkages in DNA (BOYCE, HOWARD-FLANDERS, 1964; COLE et al., 1971, 1976), double-strand breaks (CORRY, COLL, 1973), and others.

The mutations affecting individual stages of the repair process have been shown to either amplify (PARRINGTON et al., 1971; MIKHELSON et al., 1974; WITKIN, 1966) or decrease the mutation yield (WITKIN, 1969; KONDO, 1973; GEORGE, 1973). This was shown on *uvr*, *exr*, *rec* and other mutations. There are chemical modificators of replication — FdUrd, 5-aminouracil, hydroxyurea, etc. and repair processes — caffeine, acriflavine etc. (WITKIN, 1969; WITKIN, GEORGE, 1973; DRAKE, 1970; KILMAN, HARTLEY, 1968; DUBININ et al., 1972). Unspecific factors can also serve as modificators e.g. temperature, pH.

Primary lesions and mutation fixation. The nature of primary DNA lesions and the mechanism of mutation fixation may determine the modification of the mutational process. There are several types of modification mechanisms grouped according to mutagen types: 1) base analogs — bromouracil, 2-aminopurine, etc. One can amplify or weaken the mutagenic effect of these compounds by changing the efficiency of DNA polymerase functioning (DRAKE, GREENING, 1970). The same is true for modifyed bases with altered tautomeric properties which appear after exposure to nitrous acid, methylmethanesulfonate and other chemical mutagens; 2) mutagens covalently binding to DNA. On exposure to alkylating compounds there appear monoaducts, intra- and inter-strand cross-linkages, DNA-protein cross-linkages, single- and double-stranded breaks. All these lesions are nonsense mutations for the replicase. As a result, single-stranded gaps appear on the opposite strand of such lesions after replication. Repair processes in this case are similar to those resulting from ionizing radiation or UV (CLEAVER et al., 1975); 3) mutagens intercalating into DNA. This groups which includes acridines, proralenes, etidiumbromide etc. cause frame shift mutations. They appear as a result of distortion of recombination (STREISINGER et al., 1966) or replication of DNA (ROTH, 1974); 4) indirect mutagens which cause mutations through disturbances in the intracellular processes that maintain the intact structure of DNA. Inhibitors of DNA, RNA and protein synthesis can be included in this group. FdUrd has been shown to induce chromosomal aberrations (BELL, WOLF, 1964). The same applies to the effect of acidification and temperature (ZAMENHOF, 1967; FREESE, CASHEL, 1964).

Mutagens as modificators of their own effects. The idea has been first substantiated for radiation mutagenesis (DUBININ et al., 1965) and then in a number of works on chemical mutagenesis (STICH et al., 1971, 1973; SASAKI, 1973). Inactivation of repair processes by high mutagen doses leads to higher mutation yield than is theoretically expected. For a number of cases it has been shown, however, that low doses of mutagens may be much more efficient per dose unit as compared to medium and high doses (DUBININA, 1977).

Complex nature of modifications during fixation of mutations. The magnitude and direction of chemical modification is a function of many factors: mutagen and modificator dose, mutagen type, cell division stage, time period between exposure to mutagen and modificator. A wide variety of lesions may also affect the processes of repair, replication and fixation of mutations. Some stages of the mutational process may be sensibilized while others may be exposed to antimutagenic influences, etc.

In a number of cases chemical modification results in a latent state of primary lesions rather than in their repair. When *Crepis capillaris* seedlings were treated by ethylenimine on the pre-synthetic stage and subsequently exposed to 5-aminouracil — a DNA synthesis inhibitor — essentially no mutations were observed. When, however, not only the S phase but also the G_2 phase was exposed to 5-aminouracil, the mutagenic effect of ethyleneimine was restored (DUBININ et al., 1972).

Modification of chemical mutagenesis and the problem of environmental mutagens. A study of the nature of modification phenomena in chemical mutagenesis, both on exposure to individual mutagens and, especially, in the case of complex mutagenic effects, is of primary importance in studying the environmental mutagens. Both mutagens and modificators are present in the environment, antimutagenic substances are also found there (VOGEL, VRBA, 1967; OBE, SLOCIK-ERBEN, 1973).

Modifying Effects of Chemicals due to Interaction with Basic Genetic Processes

B. J. KILBEY

Institute of Animal Genetics,
West Mains Road, Edinburgh EH 9 3JN

Introduction

There is no doubt that the study of the modifying effects of chemicals on the potency of mutagenic agents has occupied a major part of the research efforts of geneticists for some years. The impetus for these studies has derived partly from a realisation that the ability to influence the outcome of radiation and chemical treatments could have far reaching effects both in medicine and plant breeding programmes [16]. However, workers have also been quick to appreciate that chemical modifiers provide us with useful tools for probing the processes of mutagenesis. This is becoming particularly important for those concerned with risk evaluation for environmental mutagens since conditions of exposure in practice are far more likely to include modifying agencies than exposures in the laboratory.

My purpose in this communication is to review, as far as I am able in the time allotted, what seem to me to be the main ways in which chemical modifiers may act. However, I think it is important to emphasize repeatedly that in studying the modification of mutagenic responses, we are embarking upon a far more extensive subject for study than we might at first think, and I should like to illustrate the point at the end of my lecture by including some of our own recent findings which reach beyond interactions *per se* and attack some more fundamental questions of metabolism and mutagenesis.

For convenience I shall divide modifying actions into three categories depending on whether their effects appear to be at (1) the level of the supply or utilisation of mutagen, (2) the interference with systems which process potential genetic lesions in the cell and [3] the expression of newly indiced but established alterations in genes. I realise of course that this is too simplistic an approach and that the divisions will inevitably become blurred.

1. Modification at the level of mutagen supply and utilization

Although modyfying effects which act by altering the permeability of the cell to a mutagen could be discussed at this point I shall not do so here. More interesting are those conditions which modify the supply of active mutagen by controlling its levels of production, or by bringing about its degradation. Until relatively recently this area of research was almost exclusively the domain of the pharmacologist. Numerous examples in the toxicological literature in which metabolic interactions have been described between drugs, insecticides, food additives, carcinogens and other environmental chemicals. Their common feature, however, is their ability to modify the metabolising capacity of liver microsomal enzymes. You will all be familiar with certain examples of this e.g. phenobarbitol-induction of microsomal systems. A complete catalogue of these effects cannot be given here, but a review of some of them may be found in publications of CONNEY and his associates [8]. Two recent reports have described examples of a similar nature in which mutation has been studied. COUCH and FRIEDMAN [9, 10] were able to show that the activity of dimethylnitrosamine in the host-mediated assay can be markedly reduced by pretreatment of the host animals (in this case Swiss male mice) with sodium nitrite, nitrosarcosine, diethylnitrosamine, dibutylnitrosamine and even dimethylnitrosamine itself. In contrast, the directly acting mutagen (β-chloroethyl)-1-nitrosourea was not subject to modification in this way and the authors therefore conclude that the pretreatments effectively reduce the activity of the liver enzyme systems concerned with the production of active mutagenic derivatives from dimethylnitrosamine. However some uncertainties remain concerning the interpretation since far more drastic treatments are known to be required for the inhibition of microsomal enzyme systems in general than were needed in these experiments to depress the mutagenic activity of dimethylnitrosamine.

Modification can operate at a level much closer to the genetic target than this however. VIG [52] has described an interesting example of synergism in human leukocytes between the drug Adriamycin, which breaks chromosomes, and deoxycytidine. Combination treatments greatly enhance chromosomal aberration frequency and the synergistic interaction appears to be a real one in that aberrations are not increased at the expense of free fragments. In this case it is suggested that, by disrupting the association between the nuclear membrane and localised heterochromatic regions, the deoxycytidine increases the number of sites available for attack by Adriamycin. This type of explanation may be more widely applicable and similar suggestions have been made to account for the potentiating effects of caffeine [18], FUdR [42] and ATP [19, 44]. Another example of modification, this time of radiation damage, which appears to operate at an early stage is the action of butylated hydroxytoluene, the food additive. As an inhibitor of the propagation of chain reactions initiated by free radicals, it might be expected to act as a radio-protective agent. In fact it is a potent radio sensitiser, enhancing the action of gamma rays in EL 4 lymphocytes [13] and in mature *Drosophila* sperm [14, 44]. The mechanism of its action is not known. However, the probable mode of action of another group of compounds, examples of which are p-nitroacetophenone and some of the nitrofurans has been described in some detail by CHAPMAN et al. [5]. These modifiers (the so-called electronaffinic sensitizers) exert their effect by fixing DNA target radicals, thus preventing their reduction and the consequent restoration of the damaged DNA strand to its initial configuration. According to these authors, the amount of strand breakage in the DNA is thus increased. In contrast to the pharmacologically-based interaction described earlier, the latter modifiers appear to act completely independently of the cellular metabolic machinery.

2. Modification at the level of repair

Biologists are obsessed by repair and most of the studies of the modifying effects of chemicals are interpreted in terms of this aspect of their activity. The one group of chemicals which has received most attention is the methylated oxypurines of which caffeine is an example. Although there is evidence of weak mutagenic activity associated with caffeine in a number of organisms, [39, 7, 12, 17] by far its most dramatic action appears to be the potentiation of the genetic effect of other agents and it appears to achieve this by the inhibition of repair processes. The repair systems affected differ in pro- and eukaryotes. As a general rule, excision repair seems to be susceptible in microbial prokaryotes [37] but it is relatively insensitive in eukaryotic systems [22]. Conversely, post replication repair in eukaryotes appears to be sensitive whether it is predominantly a recombinational type of process (as is believed to be the case for *Schizosaccharomyces pombe* by FABRE [11],) or a gap filling function which involves the de novo synthesis of DNA [29, 30]. A recent report by MCCULLEY and JOHNSON [32] indicates that post replication repair in *E. coli* B and K-12 is resistent to caffeine. This multiplicity of action may be explained in part, but only in part, if as suggested by KIHLMAN [17] caffeine interferes with repair by binding strongly to short single stranded regions in an otherwise undenatured DNA molecule. It is interesting perhaps in this context to note that a similar mechanism has been suggested to account for certain aspects of the lethal interaction between diepoxybutane and ultraviolet light in yeast [22, 23]. At present it is uncertain whether the action of caffeine on the repair systems mentioned will account for all the observed instances in which the substance potentiates the action of other mutagens. Enhancement of the mutagenic and inactivating actions of UV in prokaryotes [48, 55] is easily comprehended in these terms as the inhibition of excision repair. The potentiation of chromosomal damage produced by alkylating agents in plants and animals may, as KIHLMAN suggests, be explicable in terms of the inhibition of post replication repair. It is certainly true that in a variety of studies the maximum effects of caffeine enhancement were obtained when it was applied at the first post-treatment S. period [49, 50, 20]. However, a different mechanism may form the basis of the inhibitory effect of caffeine and actinomycin D on the rejoining of chromosome breaks induced in mature *Drosophila* spermatozoa by X-rays [36, 45]. The breaks in the chromosomes of mature sperm remain open until fertilization whereupon rapid rejoining takes place in the oocyte. Both caffeine and actinomycin D interfere with this repair, increasing the frequency of chromosomal loss and dominant lethality and reducing the frequency of translocations. The repair involved and the mechanism of its inhibition are still under examination but it has been shown that the repair system is under demonstrable genetic control [33, 34, 35].

One aspect of the inhibition of repair activity is of particular interest since it bring into sharp focus the relationship between the modifying effects of chemicals and the interpretation of dose-effect curves. This is the inhibition of cellular repair activity by the same agent which causes the DNA damage subject to repair. WARREN and STICH [54] have reported an elegant example of this from their studies of the effects of 4-nitroquinoline-1-oxide in human cells. The extent to which unscheduled DNA synthesis occurs following a measured dose of the mutagen can be markedly reduced if the cells have been pretreated with an equal dose of the same compound. The reduction in repair is at a maximum when periods of 3 hrs or less separate the conditioning dose from the main dose. The expected extent of repair is fully realised if the interval between the doses exceeds 3 hrs. Cell killing and the frequency of chromosome aberrations are enhanced when repair is depressed in this way. Other examples of split dose experiments of this type exist [eg. 51]. We have studied the interaction between diepoxybutane and UV using similar types of split dose experiment. Reversions from adenine auxotrophy to prototrophy in *Neurospora crassa* were scored in these experiments and these are probably all point mutations. Diepoxybutane enhances the sensitivity of *Neurospora* conidia to the mutagenic action of UV. We know this because the extra mutants produced by the combined treatments compared with the sum of those produced by the same treatments separately, respond to photoreactivating conditions [47]. However DEB also sensitizes conidia to further treatments with DEB itself. Many more mutants are obtained by the combination of two doses than occur by adding together the products of the individual doses. However, as with the observations of WARREN and STICH, the sensitization effect was found to disappear after about 4 hours storage of the treated conidia in water at room temperature. The decline in sensitivity in this case would seem to involve the synthesis of a new protein since it can be prevented completely by the addition of actidione, an inhibitor of protein synthesis [24] (Fig. 1). We therefore assume that the decline in sensitivity is brought about by the resynthesis of new repair enzyme molecules which replace those damaged by the first exposure to the mutagen. This

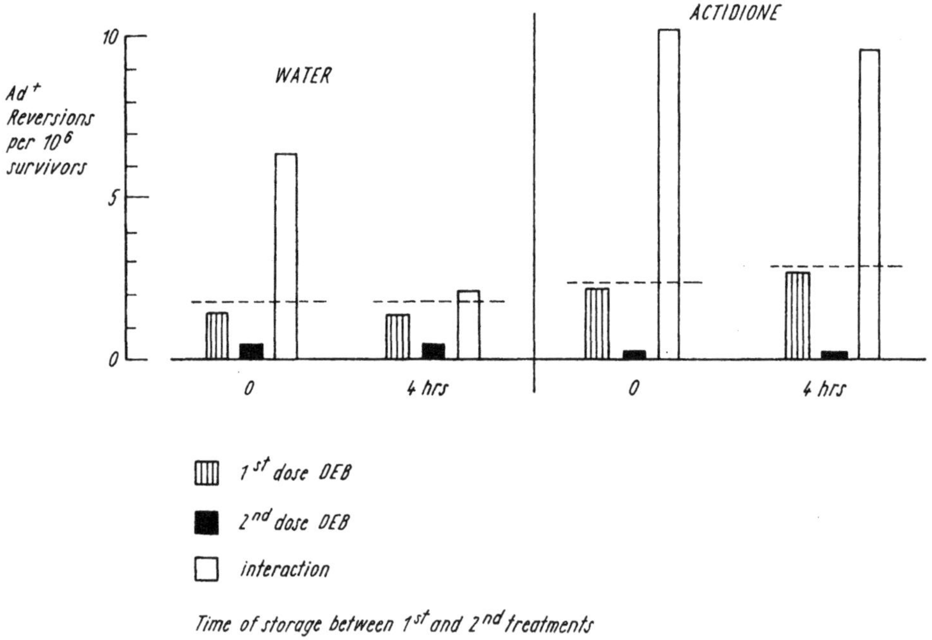

Fig. 1. The interaction between two doses of DEB in *Neurospora* when dose two follows dose one immediately or after 4 hrs incubation in water (L.H. panel) or in a solution of actidione (R.H. panel).

interpretation agrees well with the kinetic data obtained when conidia are treated with DEB at different dose-rates [28]. High dose-rates give exponential mutation induction curves presumably because the repair system is progressively inactivated and cannot be replaced quickly enough. Low dose-rates give linear mutant accumulation kinetics because damage to the repair system is neutralized by its continuous replacement. Inhibition of protein synthesis during low dose-rate treatments should, and in fact does, presumably by preventing resynthesis, convert the dose effect curve into one characteristic of high dose-rates [25]. Situations of this kind are probably very common. We have mentioned the work of WARREN and STICH in relation to human material, we have also shown similar situations exist for budding yeast [43], however in neither *Neurospora* nor in *Saccharomyces* can we yet identify the DEB-sensitive process. One way of learning more about it would clearly be to examine the spectrum of genetic damage on which it can act. To do this we have examined the induction of recessive lethality by diepoxybutane in a two component heterokaryon system first described and used by ATWOOD [1]. The recessive lethals isolated by this technique are almost certainly heterogeneous in their genetic basis ranging from point mutations in indispensible genes to deletions of several loci and possibly even whole chromosome losses. To distinguish between lethals which result from different types of genetic alteration is the main problem with using such a system and we are trying to do it by making use of another chemical modifier of mutagenic action, sodium fluoride. VOGEL [53] was recently able to show that sodium fluoride is antimutagenic when used to treat both mature and immature oocytes and mature sperm in *Drosophila* in combination with trenimon and P.D.T. Diepoxybutane also showed this interaction. Whole chromosome loss and the loss of fertility were reduced by a combination of sodium fluoride and trenimon, and a reduction in the percentage of sex-linked lethals was recorded also. The extent of the effect was, however, greatest in trenimon, less in P.D.T. and absent in the weak mutagen A 137. The data could best be explained by assuming that sodium fluoride hinders the breakage of chromosomes by these agents since the extent of the antimutagenic action parallels their known chromosome breaking ability in *Drosophila*. OBE and SLICIK-ERBEN [41] obtained support for this conclusion when they showed that the ability of trenimon and 2, 3, 5, 6 tetraethylene-imino-p-benzoquinone and A 139 to break chromosomes of human cells was also antagonised by sodium fluoride. The mechanism of the effect is obscure and it is not known whether there is any connection between it and the potentiating effects of sodium fluoride on X-ray induced recessive lethals in *Drosophila* spermatozoa [38] in barley seeds [13], or its failure to inhibit the rejoining of breaks induced by X-rays in *Drosophila* sperm, after fertilisation has occurred [33].

Miss OLSZEWSKA working in my laboratory has shown that 5 mM sodium fluoride in the plating medium has no effect on the recovery or the recovery kinetics of diepoxybutane-induced point mutations in *Neurospora* (reversions of the *ad-3 A 38701* allele), neither does it have any significant effect on the recovery of recessive lethal mutations in ATWOOD's heterokaryon system after treatment with nitrous acid. The frequency of lethals is low in this case but the two results are in agreement since BROCKMAN et al. have shown that, in the adenine 3 mutation system of *Neurospora*, nitrous acid induces very few chromosome deletions [4]. After DEB treatment, in contrast, recessive lethal induction in ATWOOD's system is drastically diminished by the presence of sodium fluoride. However, it appears that the extent of the reduction is dependent on the DEB dose used (see Fig. 2). At doses giving 5% recessive lethals or less, little or no inhibition is found but the extent of the inhibition increases rapidly as the dose is increased. Recon-

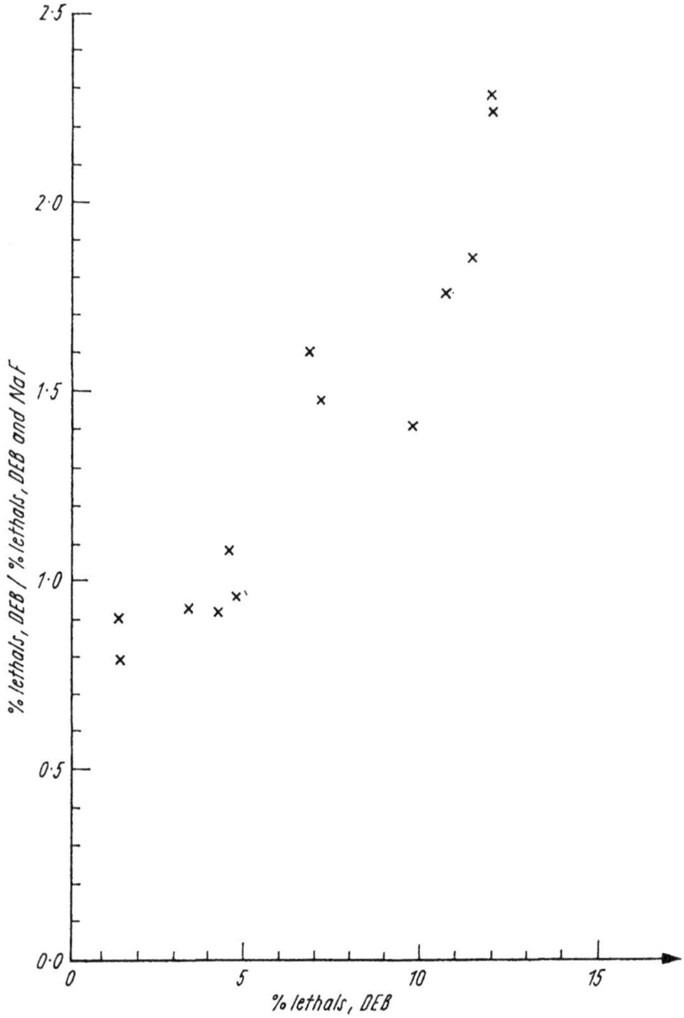

Fig. 2. The extent of the inhibitory action of NaF on the recovery of DEB induced recessive lethals in *Neurospora* after various doses of DEB (measured as % lethals induced).

struction experiments have shown that the effect of the sodium fluoride is probably exerted before the recessive lethal mutations become fixed since established recessive lethals are not susceptible to its inhibitory effects. Naturally, until we have obtained independent evidence that sodium fluoride is indeed selectively eliminating chromosome damage induced by DEB, any conclusions we reach must be very tentative. However, if for the moment this is assumed to be the case, the data show that recessive lethal mutations which arise from point mutations are not subject to the same DEB-induced enhancement as adenine reversions. And this agrees with additional observations that UV and DEB do not interact in the production of recessive lethal mutations in this system. Whether this means that the indispensible loci involved here are not subject to DEB-sensitive repair in the same way as adenine reversions or whether some other explanation applies we cannot yet say.

3. Modification at the level of expression

It is difficult to be sure that any specific examples of modification fall into this class. Methionine in the plating medium appears to interfere with the expression of adenine reversions induced by UV in *Schizosaccharomyces* [6]. Histidine appears to inhibit the expression of certain supersupressor mutations in *Saccharomyces* [46]. Diepoxybutane appears to inhibit the expression of the inositol reversions it induces itself in respiratory sufficient strains of *Neurospora*, whereas this limitation is relieved in respiratory deficient strains [3]. Unfortunately, in none of these cases is the biochemical information available to support the view that they represent interactions at the level of expression. To provide this may be difficult: Interactions at this level are unlikely to involve single, well defined cellular activities like repair systems, furthermore in many cases transient modifications of metabolism are probably responsible. However, even this difficult task may have its useful by-products since metabolic relationships could be uncovered in the analyses which were previously unsuspected. I would like to finish by summarising very briefly some work being done at present in Edinburgh by my collaborator Dr. McAthey using *Schizosaccharomyces pombe* with a view to eventually studying some of the metabolic problems which lie at the heart of our understanding of modifying agencies. Our object was to use a system in which metabolic modifications brought about by manipulations in the growth conditions could be sustained for long periods — days rather than hours or minutes, and we chose the chemostat for this purpose. The earlier bacterial workers found that, when *E. coli* is grown in the chemostat, the spontaneous accumulation of bacteriophage-resistant mutations was considerably modified by the nutrilite used to restrict the growth of the culture [40, 27]. Limitation with tryptophan led to the accumulation of mutation in proportion to absolute time, limitation with glucose or phosphate led to their accumulation in proportion to the number of divisions elapsed. The usual way of accounting for these results is to adopt the view that errors in replication are a chief source of the spontaneous mutations observed. The behaviour of cultures grown with glucose or phosphate limitation is therefore to be expected while the behaviour of cultures grown with tryptophan as the limiting nutrilite requires additional explanations. Of the explanations suggested the one most favoured by these workers appears to be that, under conditions of tryptophan limitation the rate of gene replication is constant, regardless of the rate of cellular division. Extra genomes produced at low rates of growth are discarded and one only becomes the "master nucleus" [26]. However alternative explanations based on growth rate dependent repair or expression are by no means ruled out at present. These obser-

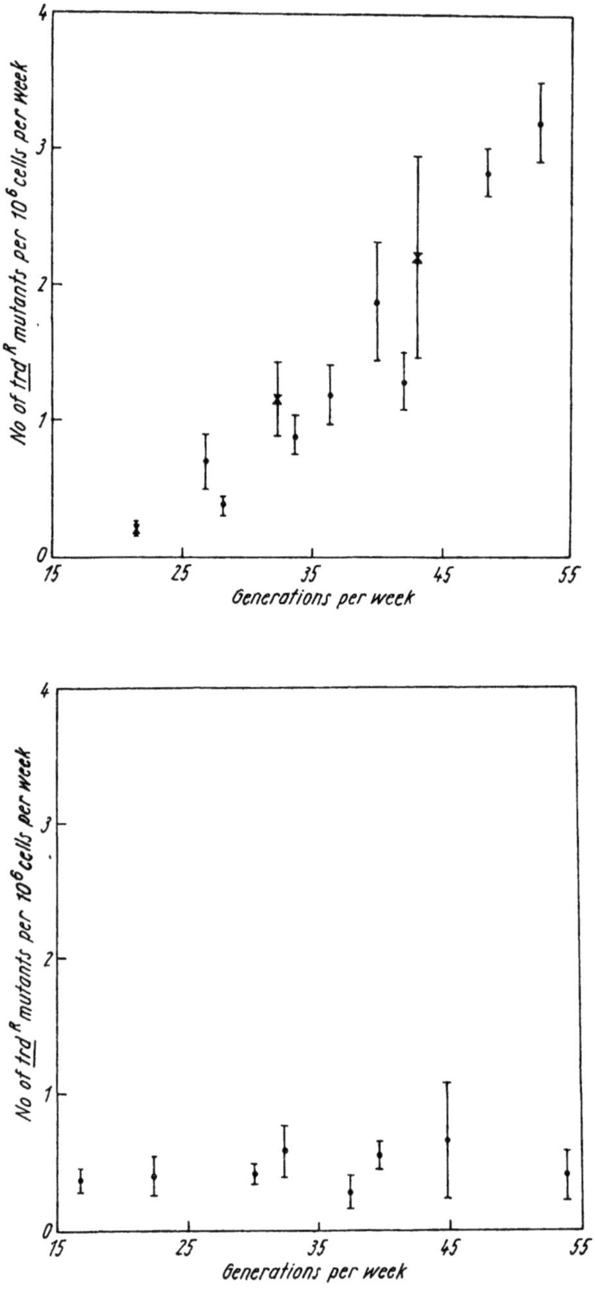

Fig. 3. The rate of accumulation of trichodermin resistant mutants in *Schizosaccharomyces pombe* as a function of generations/week when glucose (LH panel) and histidine (RH panel) are restricting growth. Each point is based on up to four experiments and the 95% confidence limits are shown by the vertical bars.

vations are obviously extremely interesting for what they might tell us about the relationship between metabolism and the control of mutagenesis in bacteria. However, our intention must always be not just to look for situations peculiar to specific systems of mutation but to try to uncover more general relationships: A result of this sort would therefore gain considerably in significance if similar behaviour could be demonstrated for, not only the *E. coli* systems referred to, but also for eukaryotic systems.

With this in mind McAthey [31] has used the mutation system trichodermin sensitivity to resistance ($trd^s \rightarrow trd^r$) to examine the effects the limiting nutrilite has on the accumulation of trd^r mutations when *Schizosaccharomyces pombe* is grown in the chemostat. So far only glucose and histidine have been used but experiments with other aminoacids are now in progress. Figure 3 shows the relationship between the mutations occurring spontaneously per unit time and the number of generations in the same period when these two nutrilites limit growth. We cannot of course, use the chemostat to cover very slow or very fast rates of growth but, over the range of generation times we can use (15—55 generations approximately per week) a surprising similarity between the bacterial and yeast systems emerges: When glucose limits growth, mutations accumulate at a rate which is dependent on the number of divisions per unit time (a week in this case) while with the amino acid histidine the rate of accumulation of mutants does not change significantly with changing numbers of divisions per week. In other words chronological time is the important variable.

The similarity between the two systems is clear but a similarity interpretation appears far less certain. Differences in patterns of DNA synthesis and cell division between these two species, eukaryote and prokaryotic, already make certain of the bacterial explanations extremely unlikely for *S. pombe*. Possibly a modifying effect based on repair or expression of new mutational damage is more correct. These conclusions will become clearer as we expand our observations to include other amino acid and small metabolites and develop other mutation systems which can be adapted to chemostat usage. We are also attempting to determine whether induced mutations fall under this control. In spite of the present uncertainties, we feel the marked similarity between the systems is an encouraging suggestion of the existence of more general relationships linking metabolism and mutagenesis. If we can understand these, we should have a firmer basis for interpreting and extending our knowledge of other more obvious modifying effects in the future.

Conclusions

From what I have said even in this very brief and incomplete survey, it is clear that the modification of mutagenic response by chemical agent is not a rare occurrence. Furthermore, although most of the investigators' attentions appear to have been centred on factors interfering with repair systems; readily comprehended and biochemically well categorized sets of functions, examples already exist which point to the importance of other levels of cellular activity as targets for interactive effects. Modifying effects may indeed present difficulties to those concerned for example with the prediction of quantitative effects of small doses of mutagen in practice when the only dose effect data available are those from well defined laboratory studies; they may well make the concept of the radequivalent a difficult one to apply in practice, but they offer us the means of investigating mutagenesis not simply as a problem of DNA chemistry but as a process involving all aspects of cellular metabolism.

Acknowledgements

I am indebted to Professor C. AUERBACH, F. R. S., Drs. Pamela McATHEY, Douglas McDONALD and Allen P. JAMES for many useful discussions. I would also like to thank Dr. McATHEY and Miss Ewa OLSZĘWSKA for allowing me to use some of their unpublished results. Work from this laboratory was supported by the Medical Research Council of the United Kingdom.

References

[1] ATWOOD, K. C. and F. MUKAI, Indispensible gene functions in *Neurospora*, Proc. Nat. Acad. Sci. Wash. **39** (1953) 1027–1035.
[2] AUERBACH, C. and D. P. RAMSEY, Analysis of a case of mutagen specificity in *Neurospora crassa* II. Interaction between treatments with diepoxybutane (DEB) and ultraviolet light. Mol. Gen. Genet. **109** (1970) 1–14.
[3] AUERBACH, C. and D. P. RAMSEY, Reversal of mutagen specificity by residual genotype, Microbial Genetics Bulletin **35** (1973) 4.
[4] BROCKMAN, H. E., F. J. de SERRES and W. E. BARNETT, Analysis of ad-3 mutants induced by nitrous acid in a heterokaryon of *Neurospora crassa*, Mutation Res. **7** (1969) 307–314.
[5] CHAPMAN, J. D., D. L. DUGLE, A. P. REUVERS, C. J. GILLESPIE and J. BORSA, Chemical radiosensitisation studies with mammalian cells growing in vitro. In: Biomedical, Chemical and Physical Perspectives. Proc. Vth Int. Congr. Radiat. Res., Seattle (1975) 752–760.
[6] CLARKE, C. H., A case of mutagen specificity attributable to a plating medium effect, Zeit. für Vererbungsl. **93** (1962) 435–440.
[7] CLARKE, C. H. and M. J. WADE, Evidence that caffeine, 8-methoxypsoralen and steroidal diamines are frameshift mutagens for *E. coli* K-12. Mutation Res. **28** (1975) 123–125.
[8] CONNEY, A. H. and J. J. BURNS, Metabolic interactions among environmental chemicals and drugs. Science **178** (1972) 576–586.
[9] COUCH, D. B., and M. A. FRIEDMAN, Suppression of dimethylnitrosamine mutagenicity by sodium nitrate. Mutation Res. **26** (1974) 371–376.
[10] COUCH, D. B., and M. A. FRIEDMAN, Suppression of dimethylnitrosamine mutagenicity by nitrosarcosine and other nitrosamines. Mutation Res. **38** (1976) 89–96.
[11] FABRE, F., Relation between repair mechanisms and induced mitotic recombination after UV irradiation, in the yeast *Schizosaccharomyces pombe*. Effects of Caffeine. Molec. Gen. Genet., **117** (1972) 153–166.
[12] FRIES, N. and B. KIHLMAN, Fungal mutations obtained with methyl zanthines, Nature, **162** (1948) 573.
[13] JAGANNATH, D. R. and C. R. BHATIA, Effect of sodium fluoride on radiation sensitivity of barley seeds, Mutation Res., **22** (1974) 51–56.
[14] KAMRA, P. O., Radio sensitizing property of butylated hydroxytoluene in *Drosophila* sperm Int. J. Radiat. Biol. **23** (1973) 295–297.
[15] KAMRA, O. P., T. GHOSE and M. MAMMEN, Radiosensitization of EL 4 lymphoma cells by a commonly used food additive – butylated hydroxytoluene (BHT), Mutation Res., **21** (1973) 293–295.
[16] KHALATKAR, A. S., A. R. GOPAL-AYEUGAR and C. R. BHATIA, Modification of mutagenic efficiency of ethyl methane sulphonate with ethidium bromide iodoacetamide and sodium fluoride in barley, Mutation Res., **27** (1975) 331–340.
[17] KIHLMAN, B. A., Effects of caffeine on the genetic material, Mutation Res. **26** (1974) 53–71.
[18] KIHLMAN, B. A., and D. KRONBORG, Caffeine, caffeine derivatives and chromosomal aberrations V. The influence of temperature and concentration on the induced aberration frequency in *Vicia faba*, Heriditas **71** (1972) 101–118.

[19] KIHLMAN, B. A., S. STURELID, K. NORALEN, and D. TIDRIKS, Caffeine, caffeine derivatives and chromosomal aberrations II. Differential response of *Allium* root tips and hamster cells to treatment with caffeine, 8-ethoxycaffeine and 6-methylcommarin, Heriditas **69** (1971) 35–50.
[20] KIHLMAN, B. A., S. STURELID, B. HARTLEY-ASP and K. NILSSON, Caffeine potentiation of the chromosome damage produced in bean root tips and Chinese Hamster cells by various chemical and physical agents, Mutation Res., **17** (1973) 271–275.
[21] KIHLMAN, B. A., S. STURELID, B. HARTLEY-ASP and K. NILSSON, The enhancement by caffeine of the frequencies of chromosomal aberrations induced in plant and animal cells by chemical and physical agents, Mutation Res., **26** (1974) 105–122.
[22] KILBEY, B. J., Diepoxybutane pretreatment effects on UV-inactivation and photo reactivation of *Saccharomyces cerevisiae*, Mutation Res., **8** (1969) 73–78.
[23] KILBEY, B. J., The interaction between diepoxybutane and ultra-violet light in yeast II. Experiments with UV-sensitive and resistant strains, Mutation Res., **10** (1970) 525–531.
[24] KILBEY, B. J., The stabilisation of a transient mutagen-sensitive state in *Neurospora* by the protein synthesis inhibitor actidione, Mol. Gen. Genet. **123** (1973) 67–72.
[25] KILBEY, B. J., The manipulation of Mutation-induction kinetics in *Neurospora crassa*, Mol. Gen. Genet., **123** (1973) 73–76.
[26] KUBITSCHEK, H. E., Intensive nuclear selection in tryptophan-limited cultures of *Escherichia coli* B/r/1, trp, Mol. Gen. Genet. **124** (1973) 269–290.
[27] KUBITSCHEK, H. E. and H. E. BENDIGKEIT, Mutation in continuous cultures. I. Dependence of mutational response upon growth-limiting factors, Mutation Res., **1** (1964) 113–120.
[28] KØLMARK, H. G. and B. J. KILBEY, Kinetic Studies of mutation induction by epoxides in *Neurospora crassa*, Mol. Gen. Genet. **101** (1968) 89–98.
[29] LEHMANN, A. R., Post replication repair of DNA in ultraviolet-irradiated mammalian cells, J. Mol. Biol. **66** (1972) 319–337.
[30] LEHMANN, A. R. and S. KIRK-BELL, Effects of caffeine and theophylline on DNA synthesis in unirradiated and UV-irradiated mammalian cells, Mutation Res., **26** (1974) 73–82.
[31] MCATHEY, P. and B. J. KILBEY, Trichodermin; a selective agent for screening forward mutations in continuous cultures of *Schizosaccharomyces pombe* Biol. Zbl. **95** (1976) 415.
[32] MCCULLEY, C. M. and R. C. JOHNSON, The absence of caffeine inhibition of post-replication repair in excision-deficient strains of *Escherichia coli* B and K-12. Mutation Res., **38** (1976) 207–214.
[33] MENDELSON, D., Lack of effect of sodium fluoride on a maternal repair system in *Drosophila* oocytes. Mutation Res., **34** (1976) 254.
[34] MENDELSON, D., The effect of caffeine on repair systems in oocytes of *Drosophila melanogaster* (I), Mutation Res., **22** (1974) 145–156.
[35] MENDELSON, D., The effect of caffeine on repair systems in oocytes of *Drosophila melanogaster* II. The induction of chromosome aberrations in irradiated males, Mutation Res. **35** (1976) 91–100.
[36] MENDELSON, D. and F. H. SOBELS, The inhibiting effect of caffeine on the maternal repair of radiation-induced chromosome breaks in *Drosophila*, Mutation Res., **26** (1974) 123–128.
[37] METZGER, K., On the dark reactivation mechanism in ultraviolet irradiated bacteria, Biochem. Biophys. Res. Commun., **15** (1964) 101–109.
[38] MUKHERJEE, R. N. and F. H. SOBELS, The effects of sodium fluoride and iodoacetamide on mutation induction by X-irradiation in mature spermatozoa of *Drosophila*, Mutation Res., **6** (1968) 217–225.
[39] NOVICK, A., Mutagens and Antimutagens, Brookhaven Symp. Biol. **8** (1956) 201–214.
[40] NOVICK, A. and L. SZILARD, Experiments with the Chemostat on spontaneous mutations of bacteria, Proc. Nat. Acad. Sci. (Wash) **43** (1950) 553–566.
[41] OBE, G., and R. SLACIK-ERBEN, Suppressive activity by fluoride on the induction of chromosome aberrations in human cells with alkylating agents in vitro. Mutation Res., **19** (1973) 369–371.
[42] OCKEY, C. H., Chromatid aberrations induced by ethyleneimines, Erwin-Bauer-Gedächtnis-Vorlesungen, I Abhandl. Deut. Akad. Wiss. Berlin, Kl. Med., I (1960) 47–53.
[43] OLSZEWSKA, E. and B. J. KILBEY, The mutagenic activity of diepoxybutane in yeast, Mutation Res., **33** (1975) 383–390.

[44] Prasad, O. and O. P. Kamra, Radiosensitization of *Drosophila* sperm by commonly used food additives — butylates hydroxyamisole and butylated hydroxytoluene, Int. J. Radiat. Bio., **25** (1974) 67—72.
[45] Proust, J. P., K. Sankaranarayanan and F. H. Sobels, The effects of treating *Drosophila* females with actinomycin-D on the yields of dominant lethals, translocations and recessive lethals recovered from X-irradiated spermatozoa, Mutation Res., **16** (1972) 65—76.
[46] Quieroz, C., The effect of the mating medium on the recovery of nonsense suppressors in *Saccharomyces cerevisiae*, Biochem. Genet., **8** (1973) 85.
[47] Rannung, U., Photoreactivation of mutational damage produced by the interaction of DEB and UV. Mol. Gen. Genet., 111 (1971) 194—196.
[48] Sauerbier, W., Inhibition of host cell reactivation in phage T 1 by caffeine, Biochem. Biophys. Res. Comm., **14** (1964) 340—346.
[49] Swietlinska, Z. and J. Zuk, Effect of caffeine on chromosome damage induced by chemical mutagens and ionising radiation in *Vicia faba* and *Secale cereale*, Mutation Res., **26** (1974) 89—97.
[50] Swietlinska, Z., D. Zaborowska and J. Zuk, Induction of chromosome aberrations by DEB and caffeine in root meristem and germinating seeds of *Vicia faba*, Mutation Res., **17** (1973) 199—205.
[51] Thilly, W. G. and C. Heidelberger, Cytotoxicity and mutagenicity of UV irradiation as a function of the interval between split doses in cultures Chinese hamster cells, Mutation Res., **17** (1973) 287—290.
[52] Vig, B. K., Synergism between deoxyribose cytidine and adriamycin in causing chromosome aberrations in human leukocytes, Mutation Res., **21** (1973) 163—170.
[53] Vogel, E., Strong antimutagenic effect of fluoride on mutation induction by trenimon and 1-phenyl-3, 3-dimethyltriazine in *Drosophila melanogaster*, Mutation Res., **20** (1973) 339—352.
[54] Warren, P. M., and H. F. Stich, Reduced DNA repair capacity and increased cytotoxicity following split doses of the mutagen 4-introquinoline-1-oxide in cultured human cells, Mutation Res., **28** (1975) 285—293.
[55] Witkin, E. M., Post-irradiation metabolism and the timing of ultraviolet-induced mutations in bacteria, Proc. 10th Intern. Cong. Genetics, Montreal, I (1958) 280—299.

Modifying Effects of Environmental Chemicals on Drug Metabolism

W. KLINGER

Institute for Pharmacology and Toxicology, DDR-69 Jena

Each test for mutagenicity comprises two components, a metabolizing and a response system. An effective metabolizing system is necessary because the majority of mutagens are primary mutagens or premutagens and require biotransformation to active metabolites or ultimate mutagens in order to mediate their effects. In some cases one may find proximal mutagens between the premutagen and the ultimate mutagen, Fig. 1.

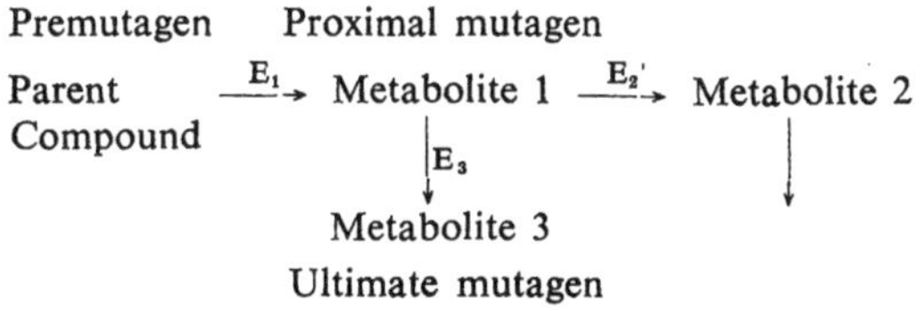

Fig. 1. Principles of biotransformation of potential premutagens. E_1, E_2, E_3 = different enzymes, metabolite 2 can be excreted, metabolite 3 acts as ultimate mutagen.

As an introductory example for this relationship may serve the well investigated mu tagenicity of 2-acetylaminofluorene, which is hydroxylated in a rate limiting cyt P-450 dependent reaction at the nitrogen to an active mutagen, which must be considered to be the proximal mutagen. The ultimate mutagen has a 20—40 fold stronger efficiency and is formed in the presence of liver fractions (FELTON et al., 1976).

The modifying effect of chemicals on drug metabolism depends on a lot of factors and it is the aim of my contribution to direct your attention to these factors, connected both with the chemical and the metabolizing system.

As a necessity these problems can be touched only to give a brief survey as a base for the following discussion. I would like to stress

First: The inhibition of biotransformation reactions by chemicals, its mechanisms and consequences,

Second: The stimulation of biotransformation reactions, its mechanisms and consequences,

Third: The interaction between inhibition and induction, exerted by the same compound, and Fourth and finally some factors influencing enzyme induction by chemicals.

I would like to start with a brief review of the processes and reactions involved in biotransformation, Fig. 2.

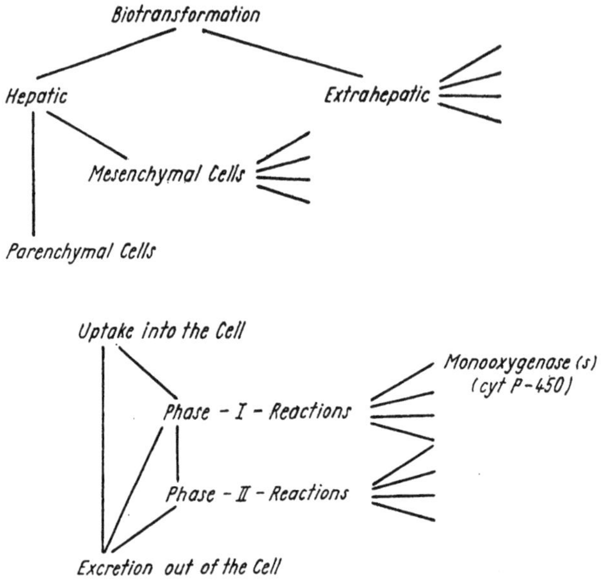

Fig. 2. Position of the liver parenchymal cell among other cells resp. organs with biotransformation activity. Steps of biotransformation within the liver parenchymal cell.

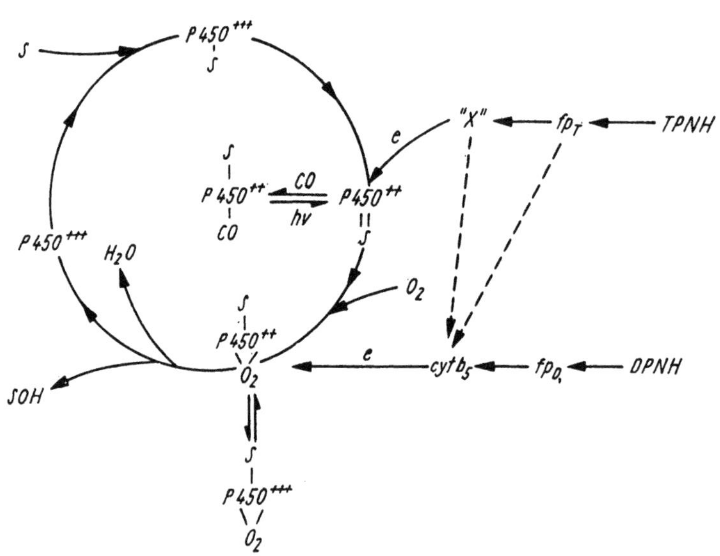

Fig. 3. Reaction scheme for the microsomal electron transport chain. The substrate to be hydroxylated is indicated by S. The flavoproteins TPNH-cytochrome P-450 reductase and DPNH-cytochrome b_5 reductase are indicated by f_{pt} and f_{pd} respectively (ESTABROOK, 1971).

Biotransformation may be hepatic (highest activities) or extrahepatic (kidney, lung, intestine).

After the uptake of the compound into the cell, in case of anions with the aid of acceptor or transport proteins, we distinguish phase I reactions (hydroxylations, reductions and hydrolysis), which result in general in reactive, more or less toxic metabolites, and phase II reactions (mainly conjugation reactions, glucuronidation, sulfatation, acetylation and so on), which result in general in water soluble non-toxic metabolites.

The most important phase I, i.e. activating or toxication reaction is the cyt P-450 dependent hydroxylation, the so-called mixed function oxidation or monooxygenation, Fig. 3.

The individual reaction steps may be briefy described as follows:

1. An organic substrate (S) reacts with the low spin form of the ferric hemoprotein to form a high spin ferric substrate complex.

2. The ferric substrate complex undergoes a one-electron reduction to a ferrous substrate complex.

3. The ferrous substrate complex reacts with oxygen to form a ternary complex of heme iron, oxygen and substrate.

4. This oxycytochrome-substrate-complex undergoes a second stage of reduction,

5. this intermediate is split setting free water and hydroxylated substrate, concomitantly the oxidized cyt p-450 is available for a new cycle.

At present the whole phase I — phase II system is understood in the following way, Fig. 4.

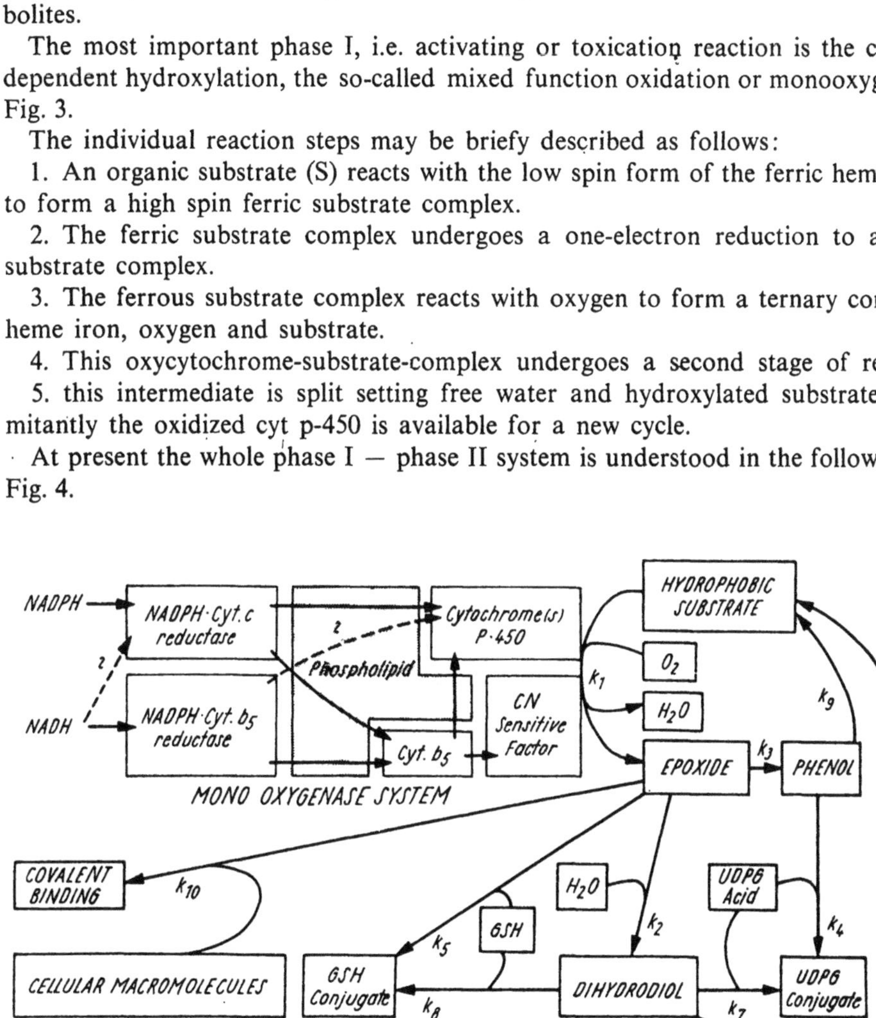

Fig. 4. Scheme for the membrane-bound multicomponent monooxygenase system(s) and the various possibly important pathways for hydrophobic substrates. For any given substrate, the relative rates of k_1 through k_{10} are currently not known and most likely differ among different tissues, strains, and species; age, nutritional, hormonal, diurnal variations, pH, and saturating versus nonsaturating conditions possibly all may be important factors in affecting these various rates (NEBERT et al., 1975).

The balance between this toxication — detoxication system, varies considerably from species to species, from strain to strain within one species, from organ and during ontogenesis.

The process of acceptor protein mediated uptake into the cell and all biotransformation reactions can be inhibited by chemicals in a competitive manner, if these compounds bind to the same binding site. But we can find allosteric inhibition as well or competitive and non-competitive inhibitions by the same compound in various concentration ranges or in depence on the enzyme source: dependent on species and age of the animals.

Some compounds are well investigated and well known inhibitors, Fig. 5, and they are used manyfold in experimental work. But we also find some often used drugs to be inhibitors, s. Fig. 6.

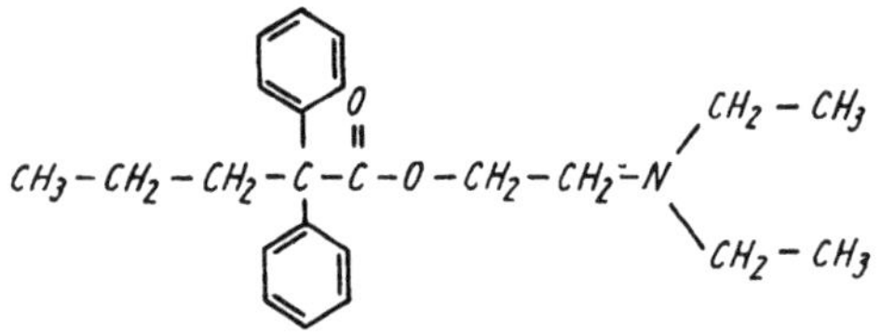

SKF 525-A

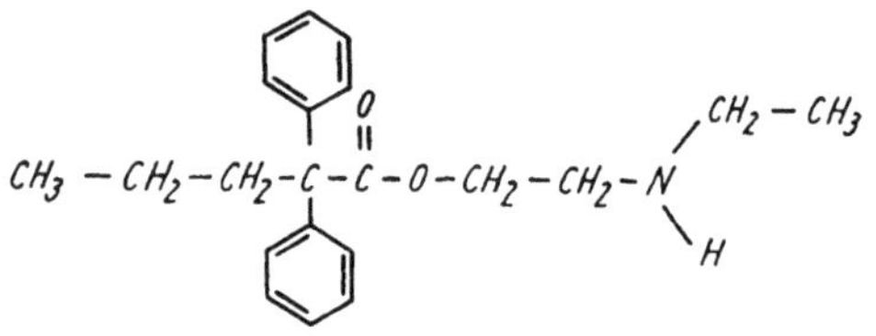

SKF 8742-A

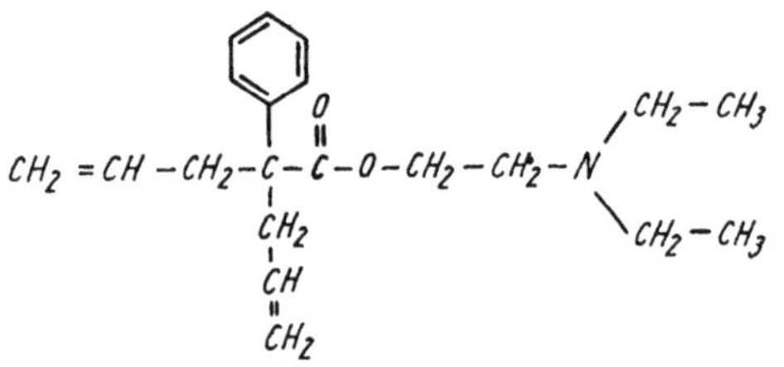

CFT 1201

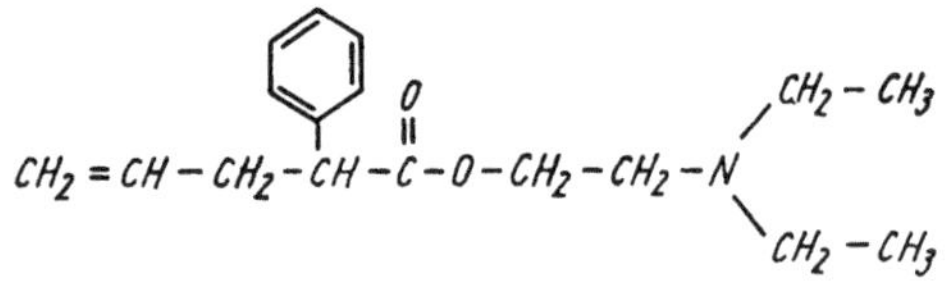

CFT 1208

Fig. 5. Structures of some strong inhibitors of microsomal monooxygenases.

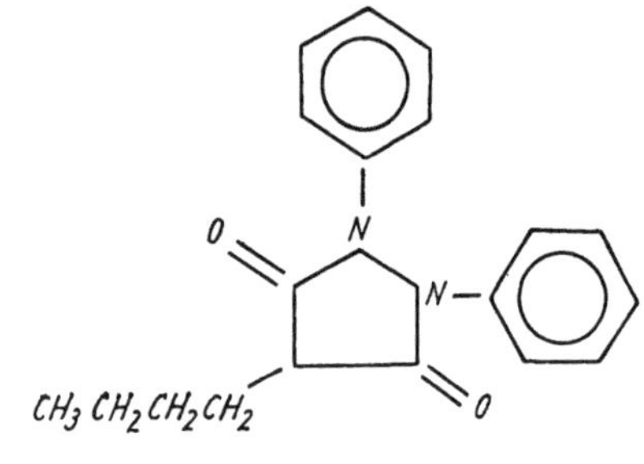

Phenylbutazone

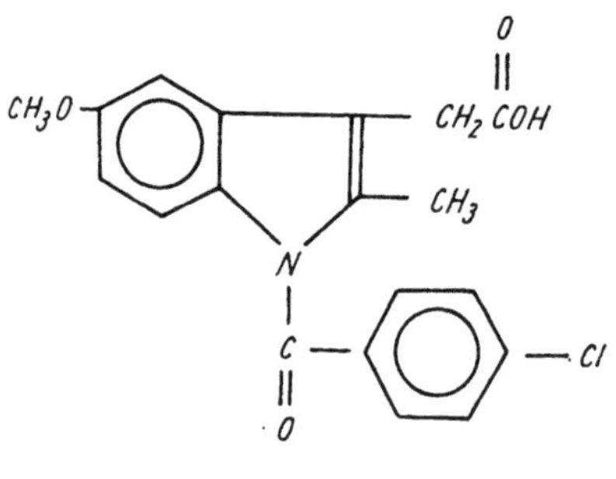

Indomethacin

Fig. 6. Structures of the antiphlogistics phenylbutazone and indomethacin.

The inhibition of one step in this reaction chain may be without consequences for parallel, forgoing or following reactions, but there may be a compensatory increase in the activity of parallel reactions. According to BELLWARD et al. (1974) the inhibition of imipramine-hydroxylation by norethindrone is followed by an compensatory increase of imipramine-N-demethylation.

In experiments with benzo(a)pyrene the inhibition of epoxydhydrase enhanced the actual concentration of the mutagenic metabolite, probably 4,5-epoxyde (RASMUSSEN and WANG, 1974).

In summary, inhibition of a toxication reaction may be followed by enhanced efficiency of another toxication or detoxication reaction. Inhibition of a detoxication reaction can be followed by increased efficiency of another detoxication reaction. For example glucuronidation can be substituted by sulfatation. But it can be followed also by an increase of the concentration of a toxication product, of activated mutagen for example. Adverse biological effects can be the consequence of biotransformation inhibition. Adverse biological effects may also be the consequence of biotransformation stimulation.

A stimulation can be observed in case of activation of inactive or latent enzyme by different mechanisms. For example aniline hydroxylation is activated in vitro by acetone, 2,2-bipyridyl, metyrapone, paroxon etc. (POWIS and BOOBIS, 1975).

The mechanism of this activation cannot be discussed here. Most interesting is the stimulation by enzyme induction. Enzyme induction leads to enhanced enzyme concentrations by increased protein de novo synthesis and/or prolongation of enzyme half life time. Inducers act on the transcriptional level by activating DNA-dependent RNA polymerases as well as on the translational level (GELBOIN 1973, HOPKINSON et al. 1974, OWENS and NEBERT, 1976) by activation of initiation factors.

Far more than 200 compounds are known to act as inducers, until now two main types of inducers: the phenobarbital and the 3-methylcholanthrene (3-MC) type of inducers can be distinguished.

They induce different types of cyt P-450 (HAUGEN et al., 1976, NEBERT and FELTON 1976) and different types of reactions. 3-MC selectively increases one cyt P-450 species, cyt P_1 450 or cyt P-448, as the absorption maximum of the reduced CO complex is at 448 nm. Moreover phenobarbital induction is accompanied by distinct augmentation of endoplasmic reticulum, i.e. microsomal protein, and liver cell hypertrophy, whereas these changes are much less pronounced in 3-MC induction. Pregnenolone-carbonitrile (PCN) and TCDD induce other types of cyt P-450. Most of the inducers belong to the phenobarbital type. Until now no structure-response-relationship has been found in this group. Within the 3-MC-group we find only polycyclic hydrocarbons, and structure-response-relationships have been described (FRANKE, 1973).

Fig. 7. Groups of drugs and other compounds which act as inducers.

Inducing Agents:

Sedatives, Hypnotics, Anesthetics, Anticonvulsants, Antihistaminics, Neuroleptics, Tranquilizers, Antidepressants, Antiphlogistics, Antidiabetics, Insecticides, Herbicides, Ingredients of Plastics, Oils, Tar, Smoke

Inducing agents we find among drugs, pesticides and other compounds, Fig. 7. Beside the differences between the main types of inducers, which have been mentioned already, the inducing power and the induction pattern differ widely.

I would like to mention only the different inducing effectiveness of various inducers for the arylhydrocarbonhydroxylase, often called benzpyren(a)hydroxylase, as an activating enzyme, and the detoxicating enzyme epoxyde hydratase (OESCH, 1976). While aryl hydrocarbon hydroxylase is inducible by 3-MC and benz(a)pyrene, epoxyde hydratase is not (BENTLEY and OESCH, 1975, LEUTZ and GELBOIN, 1975) on the other hand it could be demonstrated, that small concentrations of dieldrin as feed contamination induced epoxyde hydratase, but not aryl hydrocarbon hydratase (BELLARD, BAWSON and OTTEN, 1975). In this way the balance between toxication and detoxication enzymes was altered markedly by feed contamination. Most inducers are also potent inhibitors, and it depends on dose- and time-response-relationship, whether a chemical acts as an inducer or as an inhibitor or whether the induction effects are masked, because the inducer is present and acts as an inhibitor. But there is no general relationship, the inhibitors are inducers and vice versa.

Barbital for instance is a strong inducer of the phenobarbital-type, but it is eliminated unchanged and does not act as an inhibitor at all, phenobarbital is a moderate inhibitor, the inducer phenylbutazone is a rather strong inhibitor, Fig. 8.

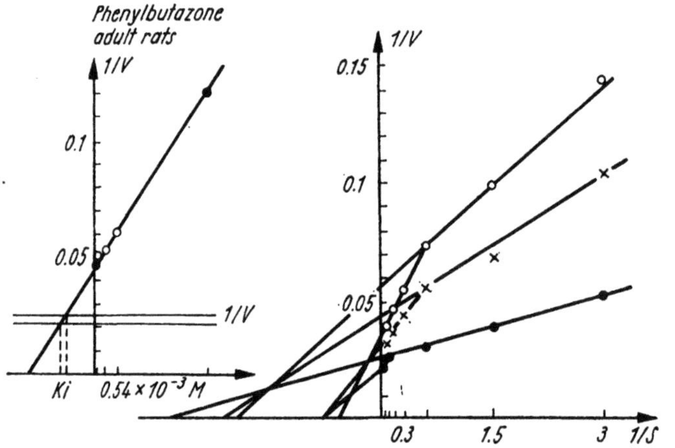

Fig. 8. The inhibitory action of phenylbutazone on amidopyrine-N-demethylation by adult rat liver 9000 g supernatant (KLINGER, 1972).

3-MC and the inducing polycyclic hydrocarbons are rather strong inhibitors of aryl hydrocarbon hydroxylase, the inducing efficiency after elimination from the animal is well correlated to the in vitro inhibitory strength (JELLINCK et al., 1975). The inhibitory effect of potent inducers of different types has been confirmed by many investigators (cp. ARCOS et al., 1975).

The biphasic action in vivo can be simply demonstrated by the following example: Immediately after administration triorthocresylphosphate inhibits the biotransformation of hexobarbital and prolongs hexobarbital sleeping time, Fig. 9. After elimination the inducing effect becomes obvious, hexobarbital breakdown is enhanced and sleeping time shortened (TRAEGER and KLINGER, 1969). More complicated are the effects of the insecticides parathion and paraoxon. Chronically administered small doses have distinct inducing effects on drug metabolism, given in high doses they act as inhibitors, but additionally they activate the metabolism of aniline (STITZEL et al., 1972). The interdependence between inhibition and induction, dose- and time-relationship and selectivity of the effects in dependence on the chemical is already rather complicated.

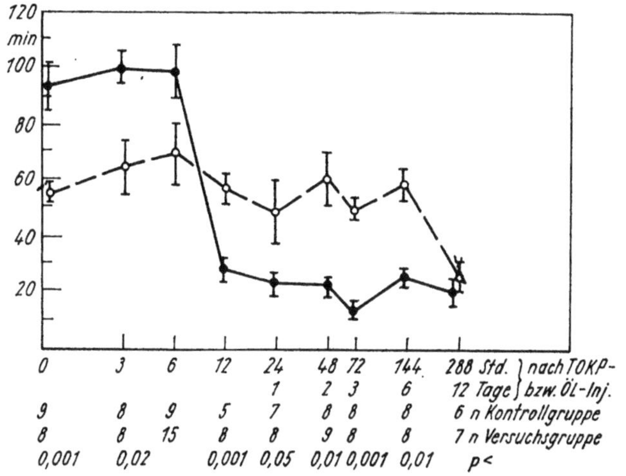

Fig. 9. Hexobarbital sleeping time (50 mg hexobarbital-Na i.p.) 0—288 h after administration of 290 mg/kg triorthocresyl phosphate (•) in 10 day-old rats. Controls received the solvent (pea nut oil) (TRAEGER and KLINGER, 1969).

It is much more intricate by the different sensitivity or responsiveness of the metabolizing system. There are very distinct species differences. For example the affinity of cyt P-450 for type I substrates differs widely (LOTNIKAR et al., 1975). On the other hand the rat is very sensitive to aflatoxin, the mouse is not. After phenobarbital pretreatment the inhibition of RNA synthesis by aflatoxin is less pronounced in rats, but increased in mice: the opposite effect in general toxicity and cancerogenicity in these two species can be demonstrated (MOULÉ et al., 1975). But even within one species basal activity of monooxygenase and its inducibility is strongly genetically determined. This has been investigated most extensively for aryl hydrocarbon hydroxylase, but also for other enzymes, (NEBERT et al., 1975, OESCH et al., 1973, BENTLEY and OESCH, 1975, HAUGEN et al., 1976, OESCH, 1976). The genetically determined responsiveness or non-responsiveness depends moreover on the inducer used: mice non-responsive against polycyclic hydrocarbons are responsive again TCDD. Further more it depends on the organ liver, lung and kidney for example (cp. ATLAS et al., 1975, HAUGEN et al., 1976).

The magnitude of organ dependent induction responses may additionally depend on the route of administration: in inhalation studies we find high aryl hydrocarbon hydroxylase activities in lung, but nevertheless the administration route is not the deci-

sive factor for the organ pattern of induction response (MARCOTTE et al., 1975, VAN CANTFORT and GIELEN, 1975). Next problem:
Especially in the rat we find after sexual maturation marked sex differences: the biotransformation activities are much higher in male than in female rats (cp. KLINGER, 1973). In all species we find low biotransformation activities in the perinatal period and a postnatal development, the velocity of which differs for the reaction type, the organ and the species under investigation (KLINGER, 1973).

Last not least the nutrional status (protein, lipids, vitamins) is of high importance for the basal activity and the inducibility, and again it is different in various organs (HAYES et al., 1973, HAYES and CAMPBELL 1974, JOLY and HÉTU, 1975, PAINE and McLEAN, 1973).

In conclusion we may say:
Chemicals can influence drug metabolism in general and mutagenicity of a compound in particular in different ways. Inhibition as well as stimulation may lead to an increase or a decrease of activation or inactivation, inhibition and stimulation can act in the same or in the opposite direction. This depends on dosage, time-response-relationship and finally on the properties of the metabolizing test system.

The same compound may act as an co-mutagen or on the contrary, it can not be classified as such, but only in connection with the defined test system.
All these facts must be taken into account
First: to avoid methodical pitfalls in testing a compound or, most difficult, a combination of chemicals for mutagenicity,
Second: for the judgement, whether the metabolic pattern of the test system is comparable to the human organism or to the possible target organ, and
Third: to evaluate not only, whether a compound must be classified as a mutagen or not, but whether a compound under certain circumstances could act as a mutagen.

References

[1] ARCOS, J. C., G. M. BRYANT, N. VENKATESAN and M. F. ARGUS, Repression of dimethylnitrosamine-demethylase by typical inducers of microsomal mixed-function oxidases, Biochem. Pharmacol. **24** (1975) 1544–1547.
[2] ATLAS, S. A., S. S. THORGEIRSON, A. R. BOOBIS, K. KUMAKI and D. W. NEBERT, Differential induction of murine Ah locus-associated monooxygenase activites in rabbit liver and kidney, Biochem. Pharmacol. **24** (1975) 2111–2116.
[3] BELLWARD, G. D., R. G. MORGAN, and V. H. SZOMBATHY, The effects of pretreatment of mice with norethindrone on the metabolism of 14 C-imipramine by the liver microsomal drug-metabolizing enzymes, Canad. J. Physiol. Pharmacol. **52** (1974) 28–38.
[4] BELLWARD, G. D., R. DAWSON and M. OTTEN, The effect of dieldrin-contaminated feed on rat hepatic microsomal epoxide hydrase and aryl hydrocarbon hydroxylase, Res. Comm. Chem. Pathol. Pharmacol. **12** (1975) 669–684.
[5] BENTLEY, P., and F. OESCH, Biosynthetic control of epoxide hydratase, Arch. pharmacol. **287** (1975) R 74
[6] VAN CANTFORT, J., and J. GIELEN, Organ specificity of aryl hydrocarbon hydroxylase induction by cigarette smoke in rats and mice, Biochem. Pharmacol. **24** (1975) 1253.
[7] FELTON, J. S., D. W. NEBERT and S. S. THORGEIRSSON, Genetic differences in 2-acetylaminofluorene mutagenicity in vitro associated with mouse hepatic aryl hydrocarbon hydroxylase activity induced by polycyclic aromatic compounds, Molec. Pharmacol. **12** (1976) 225–233.

[8] FRANKE, R., Structure-activity relationships in polycyclic aromatic hydrocarbons: induction of microsomal aryl hydrocarbon hydroxylase and its possible importance in chemical carcinogenesis, Chem. Biol. Interact. **6** (1973) 1–17.
[9] GELBOIN, H. V., 1971, Mechanisms of induction of drug metabolism enzymes, in: Concepts in Biochemical Pharmacology, Part 2, 431–451, Eds.: B. B. Brodie, J. R. Gillette, Springer Heidelberg-New York.
[10] HAUGEN, D. A., M. J. COON, and D. W. NEBERT, Induction of multiple forms of mouse liver cytochrome P-450. Evidence for gentically controlled de novo protein synthesis in response to treatment with ß-naphthoflavone or phenobarbital, J. Biol. Chem. **251** (1976) 1817–1827.
[11] HAYES, J. R., M. U. K. MGBODILE and T. C. CAMPBELL, Effect of protein deficiency of the inducibility of the hepatic microsomal drug-metabolizing enzyme system—I. effect on substrate interaction with cytochrome P-450, Biochem. Pharmacol. **22** (1973) 1005–1014.
[12] HAYES, J. R., and T. C. CAMPBELL, Effect of protein deficiency on the inducibility of the hepatic microsomal drug-metabolizing enzyme system – III. effect of 3-methylcholanthrene induction on activity and binding kinetics, Biochem. Pharmacol. **23** (1974) 1721–1731.
[13] HOPKINSON, J., P. M. PRICHARD and E. BRESNICK, Translational control at the level of initiation of protein synthesis: The effect of 3-methylcholanthrene administration on the activity of rat liver IF-M_2A, IF-M_2B and IF-M_3, Biochim. Biophys. Acta **374** (1974) 375–383.
[14] JELLINCK, P. H., G. SMITH and A. M. NEWCOMBE, Inhibition of aryl hydrocarbon hydroxylase by 3-methyl-cholanthrene, 7,8-benzoflavone and other inducers added in vitro, Chem. Biol. Interactions **11** (1975) 459–468.
[15] JOLY, J. G., and C. HÈTU, Effects of chronic ethanol administration in the rat: Relative dependency on dietary lipids – I. Induction of hepatic drug-metabolizing enzymes in vitro, Biochem. Pharmacol. **24** (1975) 1475–1480.
[16] KLINGER, W., Kinetik der Amidopyrin-N-demethylierung durch 9000 g-Leberüberstand von neugeborenen und erwachsenen Ratten in Gegenwart von Barbital, Phenobarbital und Phenylbutazon (russ.), Farmakologia i Toksikologia **35** (1972) 63–68.
[17] KLINGER, W., Biotransformation in der Leber und extrahepatische Biotransformation, in: Entwicklungspharmakologie, Ed. H. Ankermann, VEB Verlag Volk und Gesundheit Berlin 1973, p. 51–126.
[18] LEUTZ, J. C., and H. V. GELBOIN, Benzo(a)pyrene-4,5-oxide Hydratase: Assay, Properties and Induction, Arch. Biochem. Biophys. **168** (1975) 722–725.
[19] LOTNIKAR, P. D., W. J. SALDY jr., and E. N. DWYER, Dimethylnitrosamine demethylation by reconstituted liver microsomal cytochrome P-450 enzyme system, Biochem. J. **152** (1975) 705–708.
[20] MARCOTTE, J., F. S. SKELTON, M. G. COTÉ, and H. WITSCHI, Induction of aryl hydrocarbon hydroxylase in rat lung by marijuana smoke, Toxicol. Appl. Pharmacol. **33** (1975) 231–245.
[21] MOULÉ, Y., V. LESAGE, N. DARRAQ and N. ROUSSEAU, Opposite effects of phenobarbital pretreatment on aflatoxin B_1 – induced inhibition of transcription in rat and mouse liver, Biochem. Pharmacol. **24** (1975) 1851–1854.
[22] NEBERT, D. W., J. R. ROBINSON, and A. NIWA, Genetic expression of aryl hydrocarbon hydroxylase activity in the mouse, J. Cell. Physiol. **85** (1975) 393–414.
[23] NEBERT, D. S. and J. S. FELTON, Importance of genetic factors influencing the metabolism of foreign compounds, Fed. Proc. **35** (1976) 1133–1141.
[24] OESCH, F. N., MORRIS, J. W., DALY, J. E. GIELEN and D. W. NEBERT, Genetic expression of the induction of epoxide hydrase and arylhydrocarbon hydroxylase activities in the mouse by phenobarbital or 3-methylcholanthrene, Molecul. Pharmacol. **9** (1973) 692–696.
[25] OESCH, F., Differential control of rat microsomal "aryl hydrocarbon" monooxygenase and epoxide hydratase, J. Biol. Chem. **251** (1976) 79–87.
[26] OWENS, I. S., and D. W. NEBERT, Aryl hydrocarbon hydroxylase induction in mammalian liver-derived cell cultures. Effects of various metabolic inhibitors on the enzyme activity in hepatoma cells, Biochem. Pharmacol. **25** (1976) 805–813.
[27] PAINE, A. J., and A. E. M. MCLEAN, The effect of dietary protein and fat on the activity of aryl hydrocarbon hydroxylase in rat liver, kidney and lung, Biochem. Pharmacol. **22** (1973) 2875–2880.

[28] Powis, G., and A. R. Boobis, The effect of pretreating rats with 3-methylcholanthrene upon the enhancement of microsomal aniline hydroxylation by acetone and other agents, Biochem. Pharmacol. **24** (1975) 424.
[29] Rasmussen, R. E., and I. Y. Wang, Dependence of specific metabolism of benzo(a)pyrene on the inducer of hydroxylase activity, Cancer Res. **34** (1974) 2290–2295.
[30] Stitzel, R. E., J. T. Stevens and J. J. McPhillips, Effects of environmental organophosphorus insecticides on drug metabolism, Drug metabolism reviews **1** (1972) 229–249.
[31] Traeger, A. und W. Klinger, Der Einfluß von Triorthokresylphosphat auf die Hexobarbitalseitenlagenzeit verschieden alter Ratten, Acta biol. med. germ. **23** (1969) 925–928.

Drug Interactions and their Influence on Mutagenic Activity

J. Schöneich and R. Braun

Zentralinstitut für Genetik und Kulturpflanzenforschung der Akademie der Wissenschaften der DDR, DDR-4325 Gatersleben

Introduction:

The increasing demand for the genetic evaluation of chemicals in man's environment led to the establishment of mutagenicity testing systems which are being used to identify and eventually remove compounds that may cause mutations in human beings. This new field of research on public health, known as genetic toxicology [19] met greatly increased interest the last decade. In contrast to all other areas of chemical toxicology, the research on genotoxic agents is not only focused on the prevention of harmful events affecting the individual, but also his offspring [45]. Being in agreement with all other biological actions of chemical compounds on the mammalian organism, the genetic activity of chemical mutagens in the human body depends on the concentration at the receptor site in target cells, i.e. on the genetic material as well as on the time of action on it. Since such parameters are determined by basic pharmacological processes, mutation induction in mammals and man cannot be evaluated without considering the pharmacological aspects of chemical mutagenesis [41, 44, 55].

As known from clinical pharmacology [22] the bioactivity of drugs in man can be influenced by other chemical substances and by factors of environmental, pathological, and pharmacogenetic nature. In comparison with those results obtained in clinical pharmacology and experimental toxicology, little is known about the significance of drug interactions with respect to their influence on mutagenic activity. But there is no doubt that changes in mutagenic efficiency may in the same way result of drug interferences as do changes in therapeutic or toxic properties.

In this report, we would like firstly to review briefly the current state of experimental work published in the literature and obtained in our laboratory concerning the determination of mutagenic activity by pharmacological parameters, secondly we will discuss drug interactions interfering with these parameters and, in turn, modifying the mutagenic activity of chemicals in the mammalian organism, and lastly we will give a survey on drug interactions taking place outside the body or within the gastrointestinal system prior to absorption with high significance for environmental mutagenesis.

Pharmacokinetics and mutagenicity of chemical compounds:

The availability of chemical mutagens in the mammalian organism and consequently in target cells is governed by a number of principles which can be described with the pharmacokinetic behaviour of the compound (Fig. 1). Before mutation induction in

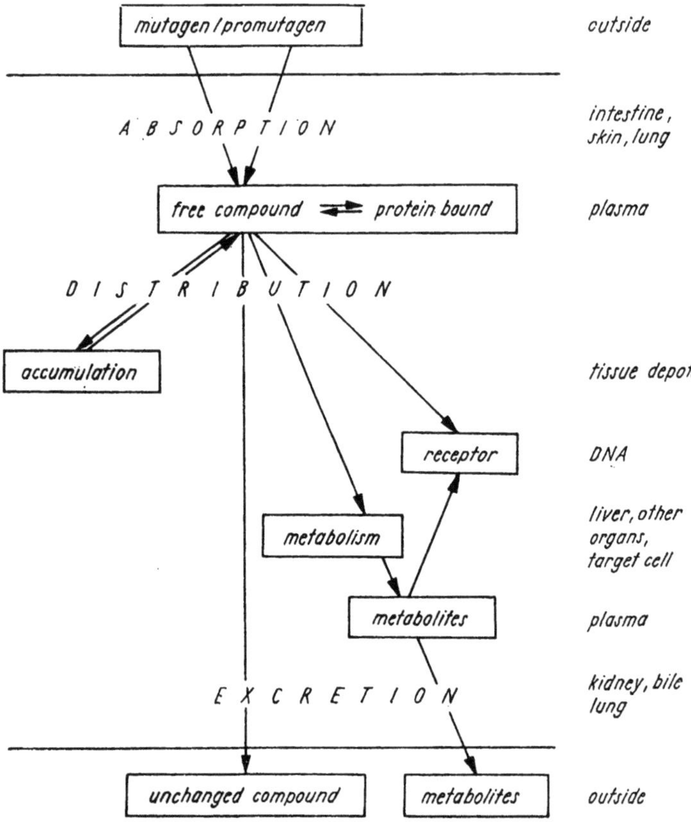

Fig. 1. Pharmacological parameters which determine the mutagenic response in mammalian organisms

body cells takes place, the compound must be absorbed, distributed through the body fluids, and metabolized in the liver and/or other tissues. Simultaneously with these processes which tend to increase the concentration of the mutagenic compounds or metabolites (invasion), the excretion via kidney, bile, or respiration will reduce the amount of the genotoxic agent and, therefore, limit its period of action on the target molecule DNA (evasion). In cases of compounds being mutagenic per se biotransformation may also contribute to a decrease of the mutagen concentration circulating in the body via inactivation. Considering these facts, mutation induction in the mammalian organism depends above all on the pharmacokinetics of the active drug species or, in other words, on the concentration of the ultimate mutagen in the body, and the time course of the body level will decide whether or not a chemical agent will be in fact mutagenic.

For example the tuberculostatic drug isoniazid should be cited. In the host-mediated assay with mice as host animals the mutagenic activity of this compound could be attributed to the metabolite hydrazine [6, 47]. The metabolic break-down of isoniazid in mice results in high concentrations of hydrazine which is present for a longer period of time in the blood stream (Fig. 2). Otherwise, in Syrian golden hamsters this metabolite shows a very different pharmacokinetic behaviour in the body and is detectable only during the first hour after the injection of isoniazid. The low excretion velocity of hydrazine from the mouse body is eventually responsible for the high mutagenic activity of isoniazid in this species, while the low mutagenic activity observed in Syrian golden hamsters can be traced back to the rapid disappearance of this metabolite.

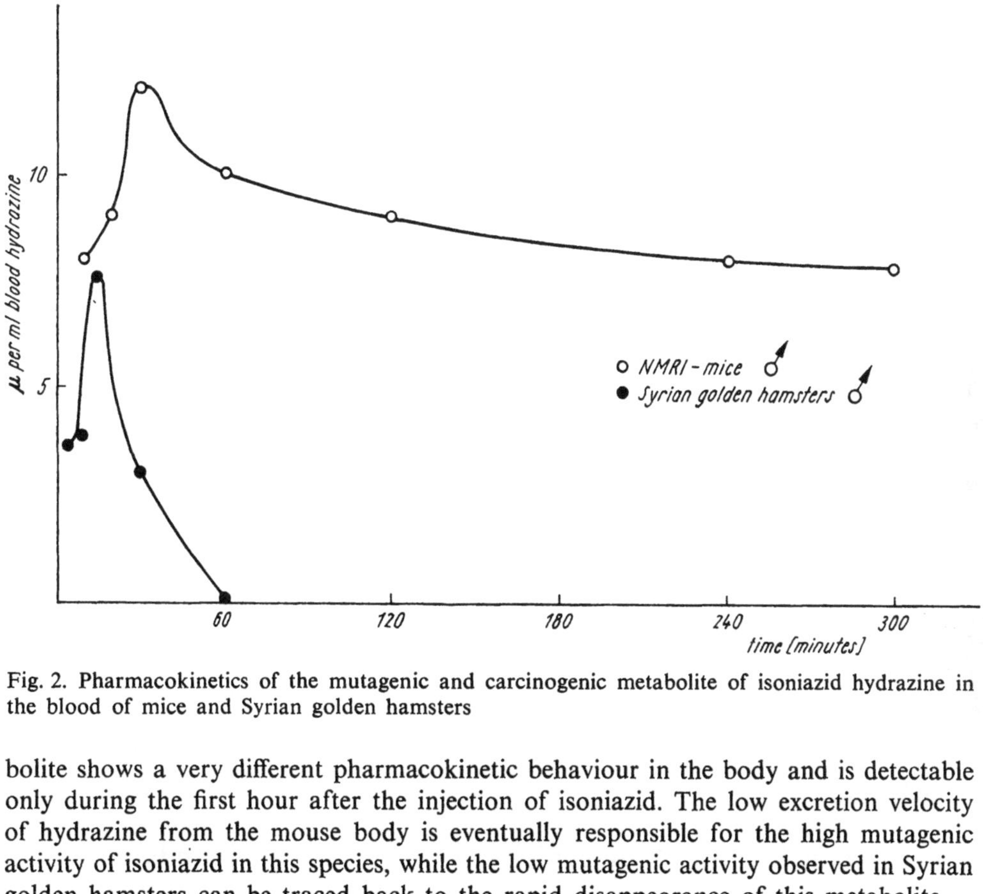

Fig. 2. Pharmacokinetics of the mutagenic and carcinogenic metabolite of isoniazid hydrazine in the blood of mice and Syrian golden hamsters

The mutagenic activity of isoniazid in various animal species, as observed in the host-mediated assay, parallelels the carcinogenicity of the compound in these species (Tab. 1). It therefore seems to be acceptable to interpret the differences in carcinogenic activity of isoniazid in various laboratory animals [3, 54] as a result of a different pharmacokinetic behaviour of the ultimate carcinogen hydrazine, as postulated by Toth and Shubik some years ago [51].

Since considerable species differences in the pharmacokinetics of a drug determine quantitative aspects of mutation induction in the mammalian organism, the estimation of blood concentrations of carcinogens and mutagens and their metabolites would be valuable in assessing the genetic risk of mutagen exposure for man. For these evaluations, the correlation of genetic activity with the dose administered and the blood levels of the ultimate mutagen that will result for the particular dose have to be determined.

Table 1

Comparison of mutagenic activity of isoniazid in the host-mediated assay with *Salmonella typhimurium* G 46 as genetic system and various laboratory animals as host organisms for the intraperitoneal culture of the bacteria with the carcinogenic activity found for isoniazid in these species

Species	Mutagenicity C-Quotient[1]	Carcinogenicity isoniazid	Carcinogenicity hydrazine	Metabolic formation of hydrazine
Mouse	9.29	+	+	+
Chinese hamster	5.05	?	?	?
Rat	4.65	+	+	+
Guinea pig	2.80	–	?	?
Syrian hamster	2.33	–	–	+

[1] C-Quotient: relative mutagenicity; Mean mutant frequency of bacteria in animals receiving isoniazid divided by the mean mutant frequency of bacteria from control animals, data from [6]. Data on carcinogenic activity of isoniazid and hydrazine from [3, 51, 54], metabolic investigations from [4, 6, 8, 54].

The pharmacokinetic behaviour estimated in animals receiving the highest non-mutagenic dose may then be used to decide whether or not exposure of man results in blood concentrations which can be characterized as safe from the point of view of mutagenicity testing with mammalian test systems. Similar deductions have recently been published for the estimation of the carcinogenic risk of chemical compounds in man [11].

Changes in the pharmacokinetic behaviour of a potential mutagen caused by other chemicals, acting either simultaneously or prior to the test compound, may exert great influences on its mutagenic activity. As known from pharmacology and toxicology, such drug interactions may take place at all steps involved in pharmacokinetics [22]. Most experimental data at hand concern interactions of chemicals with the processes involved in metabolic activation of mutagens and carcinogens, such as inhibitions or stimulations.

Drug Interactions and Mutagenicity of Chemical Compounds:

The multitude of possibilities of drug interactions and their influences on the mutagenic efficiency of chemical substances in the mammalian organism takes place, in effect, at three levels: (a) within the target cell and directly at the genetic material; (b) at the level of pharmacokinetics and biotransformation; and (c) outside the body or in the gastrointestinal system prior to the entry of the compounds into the blood stream (Fig. 3). Modifying effects due to interactions within the target cells and at the DNA level will be covered by other lectures during this symposium. We will try to give a short survey of data concerning the other levels of drug interferences mentioned under (b) and (c).

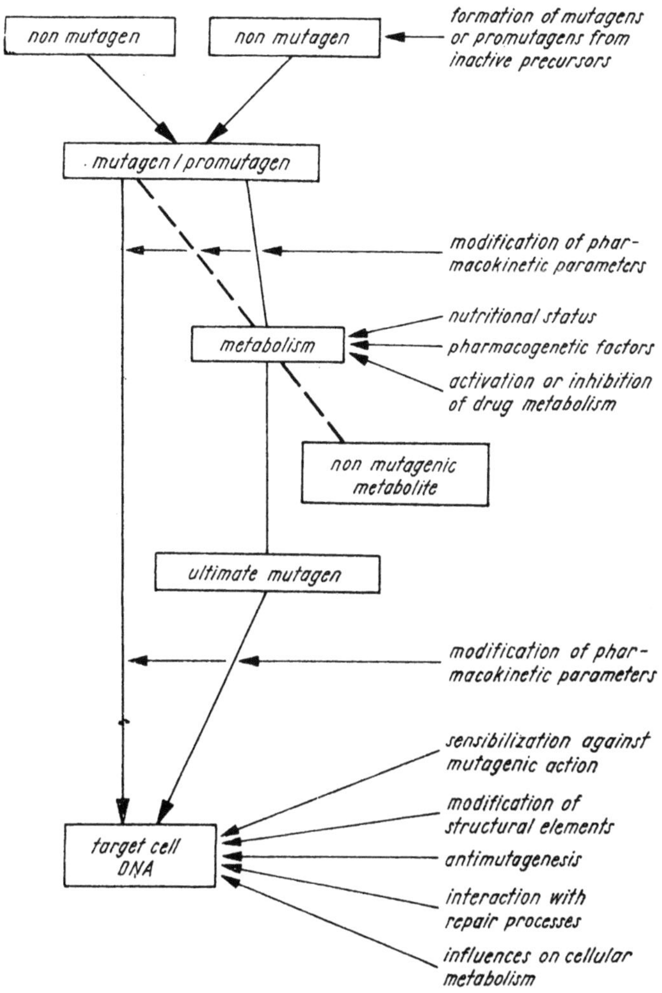

Fig. 3. Modifying effects of drug interactions on the mutagenic efficiency of chemical compounds in mammalian organisms at the level of pharmacokinetics and biotransformation, within the target cell and at the target molecule DNA, as well as outside the body or within the gastrointestinal system.

Factors influencing the metabolism and pharmacokinetics of mutagens and carcinogens:

Despite their diversity in chemical structure most carcinogens and mutagens need metabolic activation eventually resulting in the biologically active metabolite, the so-called ultimate reactive form [37]. Animal studies have shown that drug metabolism, and in the same way activation of environmental mutagens, can be influenced by many factors surveyed in part in the previous lecture given by Prof. KLINGER, such as age, sex, strain, nutrition, genetic and hormonal status, and exposure to various environmental chemicals and conditions [Ref. 2, 7, 9, 10, 32, 33, 40, 53]. The first observation on modifying effects of the nutritional status of the animals on the carcinogenic activity came from MILLER and co-workers who noted that riboflavin deficiency enhanced the carcinogenicity of aminoazo dyes [see Ref. 11]. This result has been explained by the requirement of flavin adenine dinucleotide for the enzymatic reduction of these chemicals to non-carcinogenic metabolites [39].

The modification of the genetic activity of N-nitroso compounds by the nutritional status of the animals has extensively been studied by ZEIGER using a host-mediated assay technique [58, 59]. The results obtained clearly demonstrated a lowering of the mutagenic efficiency of such promutagens in animals fed on a protein-free diet. These results have been confirmed in the in vitro microsomal assay. Hepatic microsomes from protein deficient mice possessed a lower capacity for the inactivation of the per se mutagenic N-methyl-N'-nitro-N-nitrosoguanidine and the activation of the promutagen dimethylnitrosamine [18]. An explanation for this effect may be seen in the overall reduction of the enzymatic activities following a dietary protein reduction [7] and, in consequence, a depression of metabolic formation of the ultimate reactive metabolite(s) of dimethylnitrosamine [52].

In addition, several environmental chemicals interfering with the content and activity of microsomal enzymes may cause a decrease in the metabolic turn-over of carcinogens and mutagens. As examples, organophosphorous insecticides, carbon tetrachloride, ethanol, carbon monoxide, and valerianic acid derivatives may be mentioned [10]. On the other hand, there are numerous chemicals of environmental origin stimulating the metabolism of mutagens and carcinogens in animals by inducing the de novo synthesis of cytochromes P-448 and P-450 as terminal oxidases in this process, such as polychlorinated and polybrominated hydrocarbon insecticides, urea herbicides, polycyclic aromatic compounds, dyes, and food preservatives (12). Furthermore, a great number of drugs with different chemical structure are known to be activators of biotransformations in the liver [9, 13, 31].

Table 2

The influence of activation or inhibition of biotransformation of several promutagens on their mutagenic activity as found in the host-mediated assay or bone-marrow test with mice

Test	Promutagen	Activator/Inhibitor	Influence on mutagenic efficiency	References
HMA	DMNA	Carbon tetrachloride	−	[42]
HMA	Cyclophosphamide	Carbon tetrachloride	+	[5]
Bone-marrow	Cyclophosphamide	Carbon tetrachloride	0	THIEME unpubl.
HMA	Cyclophosphamide	Ethanol	+	[5]
Bone-marrow	Cyclophosphamide	Ethanol	0	THIEME unpubl.
HMA	Cyclophosphamide	Phenobarbitone	0	BRAUN unpubl.
HMA	DMNA	Phenobarbitone	0	[42]
HMA	DMNA	Sodium nitrite	−	[14]
HMA	DMNA	Other nitroso compounds	−	[16]
HMA	DMNA	Aminoacetonitrile	−	[58]
HMA	N-Nitrosomorpholine	Aminoacetonitrile	−	[58]

+: increase in mutagenic efficiency; −: decrease; 0: no influence found.

Some of these factors have been studied with respect to their influence on mutagenic activity of chemical compounds in vitro and in vivo using microbial mammalian-mediated assays. The action of aminoacetonitrile which is known to inhibit the metabolic break-down of dimethylnitrosamine (DMNA) in the liver and, therefore, the toxic and carcinogenic manifestations [21, 27], resulted together with the promutagenic N-nitroso compounds in a decrease of the mutagenic activity of the nitroso substances in mice as compared with their activity alone (Tab. 2). Otherwise, the inhibition by aminoacetonitrile of the biotransformation into non-mutagenic forms of the per se active N-nitroso-N-methylurea was found to result in an enhancement of its mutagenic efficiency in the host-mediated assay [58].

Earlier results demonstrated a total absence on dimethylnitrosamine mutagenicity in animals with massive liver intoxications induced by carbon tetrachloride [42]. These findings are in good agreement with the demonstration that DMNA metabolism, which is an essential step towards the expression of mutagenic activity, takes place in the liver of mice [1].

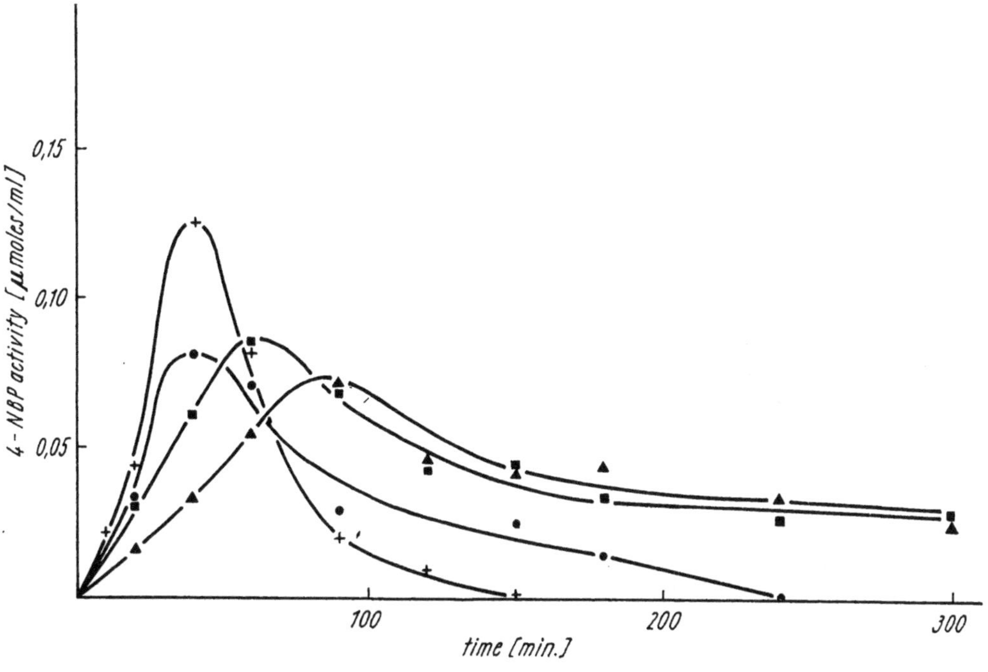

Fig. 4. Pharmacokinetic behaviour in the peritoneal cavity of alkylating break-down products produced during cyclophosphamide metabolism in carbon tetrachloride (▲), ethanol (■), or phenobarbitone (+) pretreated mice as compared with unpretreated control animals (●).

Opposite results concerning the influence of liver intoxications by carbon tetrachloride or ethanol have been reported with respect to the mutagenic efficiency of the cytostatic drug cyclophosphamide [5]. Pretreatment of mice with these chemicals increased the mutagenic activity of cyclophosphamide in the host-mediated assay as compared with results obtained from animals not pretreated. The effect of carbon tetrachloride and ethanol is due to a slowing-down of the excretion of alkylating metabolites which in turn is a consequence of a decreased activation rate in those animals. Compared to control mice, mutagenic intermediates of cyclophosphamide metabolism are present in pretreated animals over a longer period of time as demonstrated by means of pharmacokinetic analyses (Fig. 4). The total amount of alkylating products in the body as demonstrated by the area under the concentration curve (AUC; 46), is much higher under the influence of carbon tetrachloride or ethanol than under normal metabolic conditions and, therefore, responsible for the increase in mutagenic activity of cyclophosphamide in the host-mediated assay (Tab. 3).

Table 3

Comparison of mutagenic efficiency of cyclophosphamide in the host-mediated assay with *Salmonella typhimurium* TA 1950 as genetic system and mice as host animals for the intraperitoneal culture of bacteria using unpretreated animals or animals which received metabolic activators or inhibitors before injection of cyclophosphamide.

Dose Cyclophosphamide	Activator/Inhibitor and doses tested	Mutagenic activity C-Quotient	Peritoneum concentration of alkylating metabolites (AUC)	Excretion constant min^{-1}
200 mg/kg		6.59	6.753	0.01269
200 mg/kg	Carbon tetrachloride 4 ml/kg	12.42	8.018	0.00464
200 mg/kg	Ethanol 6 ml/kg	18.00	9.330	0.00557
200 mg/kg	Phenobarbitone 300 mg/kg	6.02	5.375	0.03489

AUC: Area under the concentration curve for the pharmacokinetic behaviour of alkylating metabolites in the peritoneum of mice for the first three hours after s.c. injection of 200 mg/kg cyclophosphamide (time of intraperitoneal incubation of bacteria).
Excretion constant as obtained from the statistical evaluation of the peritoneum concentration curves from several experiments.

Enzyme induction by phenobarbitone was found to be without influence on mutagenic activity in the host-mediated assay after application of dimethylnitrosamine [42] and cyclophosphamide (Tab. 3), although, phenobarbitone induces the de novo synthesis of cytochrome P-450 in liver microsomes which is an integral part of the metabolic activation system of both compounds [17, 50].

The increase in mutagenic efficiency of chemical compounds in vitro using microsomal preparations from phenobarbitone treated animals for metabolic activation is otherwise well documented for several promutagens (Tab. 4). Enzyme induction by phenobarbitone yields a higher rate of biotransformation both in vitro and in vivo, but in contrast to the living animal, the mutagenic products formed in the microsomal assay remain within a closed system as represented by the petri plate or the test tube. In living systems, however, the products of biotransformation become redistributed into the whole body and very rapidly excreted [29].

Table 4

Influence of pretreatment of animals used for the preparation of microsomal metabolic activation supply with activators or inhibitors of biotransformation on the mutagenic activity of several promutagens in the microsomal assay with bacteria as genetic indicator systems.

Activator/Inhibitor	Promutagen	Influence on mutagenic efficiency	References
Phenobarbitone	DMNA	+	[23, 26]
	Vinyl chloride	+	[36]
	DENA	+	[1]
	Aflatoxin B_1	+	[24]
3-Methylcholanthrene	DMNA	+	[26, 23]
	Aflatoxin B_1	0	[24]
Arochlor 1254[1]	DMNA	+	[26]
	Vinyl chloride	0	[25]
Vinyl chloride	DMNA	–	[25]
	Vinyl chloride	0	[25]
Benzo(a)pyrene	DMNA	–	[26]
SKF 525 A[2]	DMNA	–	[26]
CO[3]	DMNA	–	[17]

+: increase in mutagenic efficiency; –: decrease; 0: no influence found.
[1] Arochlor 1254: Mixture from polychlorinated biphenyls.
[2] SKF 525 A: 2-diethylaminoethyl-2,2-diphenylvalerate.
[3] CO: Carbon monoxide.

Results obtained in the in vitro system reflect the circumstances of metabolic activation, but they cannot explain the consequences for the living mammalian organism. Pretreatment of mice or rats which in turn have been used for the preparation of the microsomal activation supply with phenobarbitone or 3-methylcholanthrene resulted in a significant increase in mutagenic activity of N-nitrosamines in the microsomal assay [1, 23]. These findings are in good conformity with the increase in the activity of microsomal dealkylases in pretreated animals which are integrated in the metabolic formation of the ultimate mutagens and carcinogens [35.].

Attempts to increase dialkylnitrosamine carcinogenesis with phenobarbitone or 3-methylcholanthrene yielded unexpected results. Treatment of mice or rats with diethylnitrosamine (DENA) and phenobarbitone decreased the incidence of tumors in these animals as compared with the activity of the N-nitroso compound in control animals [30]. The induction of microsomal hydroxylases also protected rodents from the carcinogenic effects of other unrelated carcinogens all of which need metabolic activation [for a review see Ref. 11].

All these studies suggest that both stimulation and inhibition of microsomal metabolism can, under appropriate conditions, inhibit the mutagenic and carcinogenic activity of chemicals in the mammalian organism. The outcomes of drug interactions at the site of metabolism affecting the mutagenic activity in mammals can only be interpreted if the pharmacokinetic behaviour of the ultimate mutagen under the influence of the modifying agent is taken into consideration.

No investigations concerning interactions with other phases of pharmacokinetics, such as absorption, distribution, and excretion are available, although such interactions take important influences on the mutagenic efficiency of chemical compounds. It should be mentioned here that pathological changes of these processes may be expected to contribute to an increase or decrease in mutagenic activity.

Genetic hazards due to N-nitroso compound formation:

Further important drug interactions with high significance from the point of view of a mutagenic hazard for man are examplified by N-nitrosation reactions which may take place outside the body or within the gastrointestinal system prior to absorption and may yield mutagenic and carcinogenic N-nitroso compounds from inactive precursors. Such nitrosatable chemicals are secondary and tertiary amines, quarternary ammonium compounds, ureas, carbamates and guanidines most of them are widely distributed within the environment and may be drugs, pesticides, food constituents, and industrial chemicals. About 30 amines and amides have so far been tested for long term carcinogenicity together with nitrite the main source for nitrosating agents.

First attempts to demonstrate the mutagenic potential of N-nitroso compounds formed from inactive precursors were based on in vitro studies [28]. In vivo studies have later been published [15, 49]. For the demonstration of the mutagenic activity of amine or amide-nitrite preparations the host-mediated assay using microbial indicators seems to be the most suitable system today available. This is due to the acid catalyzed reaction within the mammalian stomach [48] and the predominant induction of point mutations by the N-nitroso compounds formed [56, 57].

Evidence has been obtained that the carcinogenic and mutagenic hazard of these interactions is preferentially determined by the turn-over rate of the amine or amide into the N-nitroso compound. Thus, the low molecular dialkylamines, whose N-nitroso

Table 5

Comparison of the mutagenic and carcinogenic activity of sodium nitrite orally applicated to laboratory animals in combination with secondary amines or amides

Compound	Carcinogenesis Rat (strain)	Mouse (strain)	Mutagenesis HMA (mouse strain)	References
Dimethylamine	SIV –	A –	ICR +	[15]
			NMRI –	Braun unpubl.[1]
Methylbenzalamine	SIV ++	NMRI ++	NMRI –	Braun unpubl.
Morpholine	SIV ++++	A ++++	NMRI +++	Braun unpubl.
Piperazine	SIV lethal	A ++++	NMRI ++++	Braun unpubl.
Aminotriazole	?	?	NMRI +	Braun unpubl.
Ethylurea	SIV +++	A ++++	ICR ++	[15]
			NMRI +	Braun unpubl.
Methylurea	SIV ++++	A ++++	ICR ++++	[15]
			NMRI +++	Braun unpubl.
Ethylthiourea	?	?	NMRI ++++	Braun unpubl.

Results obtained for carcinogenic activity from [38, 48].

derivatives are known to be strong carcinogens and mutagens [4, 20, 34], are without any significant mutagenic and carcinogenic potential when given together with nitrite to mice or rats (Tab. 5). The cyclic amines morpholine and piperazine and all urea derivatives tested together with nitrite were found to yield significant amounts of N-nitroso compounds detectable in the host-mediated assay. Especially for piparazine and ethylenethiourea a very high turn-over rate into the mutagenic N-nitroso derivatives has been found. These results obtained in our laboratory are in conformity with those obtained from studies in experimental oncology [48]. One should not forget that, however, as opposed to long term studies for the demonstration of the carcinogenic risk, in the mutagenicity test system used only the nitrosation product of a single very high dose can be estimated. Consequently, mutagenic activity can only be demonstrated for such amines or amides in the presence of nitrite which are transformed into N-nitroso compounds with high intensity and whose nitroso derivative represent a strong mutagen. Further research is needed to test the genetic activity of lower doses which correspond to the real exposure of man over a longer period of time.

Piperazine and morpholine, which have been demonstrated to yield mutagenic nitrosamines, are still used as drugs or drug constituents and may come in close contact with nitrite from environmental origin in the human stomach to give optimal conditions for N-nitroso compound formation [38]. They, therefore, represent a genetic hazard for man, although, these compounds themselves are not mutagenic.

These results may serve as examples to demonstrate drug interactions which transform a potential mutagen into a mutagenic compound. It is thus impossible to evaluate the genetic activity of chemical compounds in man's environment without considering such important sources of mutagenic hazards.

Comment:

What conclusions can be drawn from all these investigations concerning the modifying effects of drug interactions on the mutagenic efficiency of chemical compounds in the mammalian organism?

First of all, we must be aware that the mutagenic activity of chemicals in man may differ quantitatively from results obtained in routine mutagenicity testing, because the mutagenic effectivity of chemicals not only depends on their reactivity at the DNA level but also on the individuals physiological state and on presence of other compounds. As demonstrated above, such interactions between chemicals may be beneficial or hazardous for the individual from the point of view of genetic toxicology. In addition, the results of such drug interactions may vary from species to species and from person to person, and can be of great clinical significance or without any practical consequences.

A modern trend in medicine is the simultaneous treatment of patients with two or more drugs. For only one day hospital curing the administration of ten and more drugs is not a rare exception [46]. Numerous therapeutics contain two and more active ingredients or contain, in addition to the active species, constituents without therapeutic significance. Furthermore, patients may abuse during therapeutic managements drugs from irregular sources in addition to those prescribed. Considering these facts a drug interaction cannot be excluded and may yield a mutagenic activity unexpected from those obtained in mutagenicity testing with mammals.

A second aspect which should not be underestimated in attempts to evaluate the mutagenic potential of chemicals for man is the exposure to environmental substances and poisons and their accumulation in body tissues. As examplified by chlorinated hydrocarbon pesticides, the accumulated chemicals may take influence on pharmacological processes for long periods of time after their uptake. This phenomenon is without doubt of great interest for the evaluation of the genetic risk for subpopulations, such as workers in the chemical industry, agriculture, and other chemically contaminated working places.

The experimental data reported and surveyed here should be understood as a first starting point to demonstrate that drug interactions may take great influences on the mutagenic efficiency of chemical compounds in mammalian organisms. Further research is, however, needed to obtain a more comprehensive picture of their significance with respect to the mutagenic activity of drugs and environmental chemicals in man. We must recognize that mutagenic activity in human beings cannot be sufficiently described on the basis of data obtained in mutagenicity testing systems, although their use is clearly the first essential step for evaluating genetic risks in man. Moreover, mutation induction in mammals and man is a dynamic process which we can fully understand only if all factors possibly affecting this process are taken into consideration. Among these drug interactions are without doubt of much importance.

References

[1] Bartsch, H., C. Malaveille and R. Montesano, In vitro metabolism and microsome-mediated mutagenicity of dialkylnitrosamines in rat, hamster, and mouse tissues, Cancer Res., **35** (1975) 644–651.

[2] Beliles, R. P., The influence of pregnancy on the acute toxicity of various compounds in mice, Toxicol. Appl. Pharmacol., **23** (1972) 537–540.

[3] Biancifiori, C. and L. Severi, The relation of isoniazid (INH) and allied compounds to carcinogenesis in some species of small laboratory animals: a review, Brit. J. Cancer, **20** (1966) 528–538.

[4] Boxenbaum, H.-G. and S. Riegelman, Determination of isoniazid in biological fluids, J. Pharm. Sci., **63** (1974) 1191–1197.

[5] Braun, R. and J. Schöneich, The influence of ethanol and carbon tetrachloride on the mutagenic effectivity of cyclophosphamide in the host-mediated assay with *Salmonella typhimurium* Mutation Res., **31** (1975) 191–194.

[6] Braun, R., J. Schubert and J. Schöneich, On the mutagenicity of isoniazid, Biol. Zbl., **95** (1976) 423–436.

[7] Campbell, T. C. and J. R. Hayes, Role of nutrition in the drug-metabolizing enzyme system, Pharmacol. Revs., **26** (1974) 171–197.

[8] Colvin, L. B., Metabolic fate of hydrazines and hydrazides, J. Pharm. Sci., **58** (1969) 1433–1443.

[9] Conney, A. H. and J. J. Burns, Factors influencing drug metabolism, Advanc. Pharmacol., **1** (1962) 31–58.

[10] Conney, A. H. and J. J. Burns, Metabolic interactions among environmental chemicals and drugs, Science, **178** (1972) 576–586.

[11] Conney, A. H. and W. Levin, Carcinogen metabolism in experimental animals and man, IARC Scientific Publ., **10** (1974) 3–22.

[12] Conney, A. H., Environmental factors influencing drug metabolism, in: Ladu, B. N., Mandel, H. G. and Way, E. L., Fundamentals in drug metabolism and drug disposition, Williams & Wilkins Co., Baltimore (1971) 253–278.

[13] CONNEY, A. H., Pharmacological implications of microsomal enzyme induction, Pharmacol. Revs., **19** (1967) 317–366.
[14] COUCH, D. B. and M. A. FRIEDMAN, Suppression of dimethylnitrosamine mutagenicity by sodium nitrite, Mutation Res., **26** (1974) 371–376.
[15] COUCH, D. B. and M. A. FRIEDMAN, Interactive mutagenicity of sodium nitrite, dimethylamine, methylurea, and ethylurea, Mutation Res., **31** (1975) 109–114.
[16] COUCH, D. B. and M. A. FRIEDMAN, Suppression of dimethylnitrosamine mutagenicity by nitrososarcosin and other nitrosamines, Mutation Res., **38** (1976) 89–96.
[17] CZYGAN, P., H. GREIM, A. J. GARRO, F. HUTTERER, F. SHAFNER, H. PROPPER, O. ROSENTHAL and D. Y. COOPER, Microsomal metabolism of dimethylnitrosamine and the cytochrome P-450 depency of its activation to a mutagen, Cancer Res., **33** (1973) 2983–2986.
[18] CZYGAN, P., H. GREIM, A. J. GARRO, F. SHAFNER, and H. PROPPER, The effect of dietary protein-deficiency on the ability of isolated hepatic microsomes to alter the mutagenicity of a primary and a secondary carcinogen, Cancer Res., **34** (1974) 119–123.
[19] DE SERRES, F. J., Prospects for a revolution in the methods of toxicological evaluation, Mutation Res., **38** (1976) 165–176.
[20] DRUCKREY, H., R. PREUSSMANN, S. IVANKOVIC and D. SCHMÄHL, Organotrope carcinogene Wirkung bei 65 verschiedenen N-Nitroso-Verbindungen an BD-Ratten, Z. Krebsforsch., **69** (1967) 103–201.
[21] FIUME, L. and S. ROFFIA, Inhibition of aminoacetonitrile of dimethylnitrosamine metabolism in the rat's liver, Nature, **206** (1973) 1157–1158.
[22] FORTH, W., Arzneimittelinterferenz (drug-drug interactions), Arzneim.-Forsch., **26** (1976) 108–114.
[23] FRANTZ, C. N. and H. V. MALLING, Factors affecting metabolism and mutagenicity of dimethylnitrosamine and diethylnitrosamine, Cancer Res., **35** (1975) 2307–2314.
[24] GARNER, R. C. and C. M. WRIGHT, Induction of mutations in bacteria by liver microsomal metabolites of aflatoxin B_1, Brit. J. Cancer, **28** (1973) 544–551.
[25] GARRO, A. J., J. B. GUTTENPLAN and P. MILVY, Vinyl chloride dependent mutagenesis: effects of liver extracts and free radicals, Mutation Res., **38** (1976) 81–88.
[26] GUTTENPLAN, J. B., F. HUTTERER and A. J. GARRO, Effects of cytochrome P-448 and P-450 inducers on microsomal dimethylnitrosamine demethylase activity and the capacity of isolated microsomes to activate dimethylnitrosamine to a mutagen, Mutation Res., **35** (1976) 415–422.
[27] HADJIOLOV, D., The inhibition of dimethylnitrosamine carcinogenesis in rat liver by aminoacetonitrile, Z. Krebsforsch., **76** (1971) 91–92.
[28] HUSSAIN, S. and L. EHRENBERG, Mutagenicity of primary amines combined with nitrite, Mutation Res., **26** (1974) 419–422.
[29] KAPPAS, A. and A. P. ALVARES, How the liver metabolize foreign substances, Scientific American, **232** (1975) 22–31.
[30] KUNZ, W., G. SCHAUDE and C. THOMAS, Die Beeinflussung der Nitrosamincarcinogenese durch Phenobarbital und Halogenkohlenwasserstoffe, Z. Krebsforsch., **72** (1969) 291–304.
[31] KUNTZMAN, R., Drugs and enzyme induction, Ann. Rev. Pharmacol., **9** (1969) 21–36.
[32] LADU, B. N., Genetic factors modifying drug metabolism and drug response, in: Ladu, B. N., Mandel, H. G. and Way, E. L., Fundamentals in drug metabolism and drug disposition, Williams & Wilkins Co., Baltimore (1971) 308–327.
[33] MACLEOD, S. M., K. W. RENTON and N. R. EADE, Development of hepatic microsomal drug-oxidizing enzymes in immature male and female rats, J. Pharmacol. exptl. Therap., **183** (1972) 489–498.
[34] MAGEE, P. N. and J. M. BARNES, Carcinogenic nitroso compounds, Advan. Cancer Res., **10** (1967) 163–246.
[35] MAGOUR, S. and J. G. NIEVEL, Effect of inducers of drug-metabolizing enzymes on diethylnitrosamine metabolism and toxicity, Biochem. J., **123** (1971) 8p–9p.
[36] MALAVEILLE, C., H. BARTSCH, A. BARBIN, A. M. CAMUS and R. MONTESANO, Mutagenicity of vinyl chloride, chloroethyleneoxide, chloroacetaldehyde and chloroethanol, Biochem. Biophys. Res. Communic., **63** (1975) 363–370.

[37] MILLER, J. A., Carcinogenesis by chemicals: An overview-G.H.A. Clowes Memorial Lecture, Cancer Res., **30** (1970) 559–576.

[38] MIRVISH, S. S., Formation of N-nitroso compounds: Chemistry, kinetics, and in vivo occurence, Toxicol. Appl. Pharmacol., **31** (1975) 325–351.

[39] MUELLER, G. C. and J. A. MILLER, The metabolism of 4-dimethylaminoazobenzene by rat liver homogenates, J. Biol. Chem., **176** (1948) 535–544.

[40] NEBERT, D. W. and J. R. ROBINSON, Genetic and environmental factors influencing drug metabolism, in: Morselli, P. L., Garattini, S. and Cohen, S. N., Drug interactions, Raven Press, New York (1974) 51–63.

[41] NEUBERT, D., The toxicological evaluation of mutagenic events, Mutation Res., **25** (1974) 145–157.

[42] PROPPING, P. and W. BUSELMAIER, The influence of metabolism on mutagenic activity in the host-mediated assay, Arch. Toxicol., **28** (1971) 129–134.

[43] PROPPING, P., W. BUSELMAIER und G. RÖHRBORN, Kritische Betrachtungen über die intraanimale Kultur von Mikroorganismen, eine Methode zum Nachweis chemisch induzierter Mutationen, Arneim.-Forsch., **23** (1973) 746–749.

[44] RALL, D. P., Pharmacokinetic and other factors related to mutagenicity testing: Quantitative analysis of the testing procedure, Environm. Hlth. Perspec., **6** (1973) 67–70.

[45] RAMEL, C. (Ed.), Evaluation of genetic risk of environmental chemicals, Ambio Special Report, **3** (1973) 1–27.

[46] RITSCHEL, W. A., Angewandte Biopharmazie, Wissenschaftliche Verlagsgesellschaft mbH, Stuttgart (1973).

[47] RÖHRBORN, G., P. PROPPING and W. BUSELMAIER, Mutagenic activity of isoniazid and hydrazine in mammalian test systems, Mutation Res., **16** (1972) 189–194.

[48] SANDER, J. and F. SCHWEINSBERG, Wechselbeziehungen zwischen Nitrat, Nitrit und kanzerogenen N-Nitrosoverbindungen, 1. und 2. Mitteilung, Zbl. Bakt. Hyg., I. Abt. Orig. B, **156** (1972) 299–340.

[49] SEILER, J. P., In vivo mutagenic interaction of nitrite and ethylenethiourea, Experientia, **31** (1975) 214–215.

[50] SLADEK, N. E., Metabolism of cyclophosphamide by rat hepatic microsomes, Cancer Res. **31** (1971) 901–908.

[51] TOTH, B. and P. SHUBIK, Lack of carcinogenic effects of isonicotinic acid hydrazide in Syrian golden hamsters, Tumori, **55** (1969) 127–136.

[52] VENKATESAN, N., M. F. ARGUS and J. C. ARCOS, Mechanism of 3-methylcholanthrene-induced inhibition of dimethylnitrosamine demethylase in rat liver, Cancer Res., **30** (1970) 2556–2562.

[53] VESELL, E. S., C. M. LANG, W. J. WHITE, G. T. PASSANANTI and S. L. TRIPP, Hepatic drug metabolism in rats: Impairment in a dirty environment, Science, **179** (1973) 896–897.

[54] VIVIEN, J. N., R. THIBIER and A. LEPEUPLE, Recent studies on isoniazid, Adv. Tub. Res., **18** (1972) 148–230.

[55] VOGEL, F., G. RÖHRBORN and I. HANSMANN, Die Testung von Fremdstoffen auf Mutagenität, Arzneim.-Forsch., **24** (1974) 1665–1677.

[56] VOGEL, E., The relation between mutational pattern and concentration of chemical mutagens in *Drosophila*, IARC Scientific Publ., **12** (1976) 117–137.

[57] VOGEL, E. and F. H. SOBELS, Mutagenitätsprüfung an *Drosophila* als Instrument zur Erfassung potentieller Carcinogene, Biol. Zbl., **95** (1976) 405–413.

[58] ZEIGER, E., Some factors affecting the host-mediated assay response, Environm. Hlth. Perspec., **6** (1973) 101–109.

[59] ZEIGER, E., Dietary modifications affecting the mutagenicity of N-nitroso compounds in the host-mediated assay, Cancer Res., **35** (1975) 1813–1818.

Antimutagenesis and Antimutagens

C. H. Clarke

School of Biological Sciences, Iniversity of East Anglia, Norwich NR 4 7 TJ (England)

Summary

This review deals with aspects of antimutagenesis not covered in an earlier paper by Clarke and Shankel [2]. The topics now mentioned are:

a) Antimutagenesis by analogues and amino mixtures against NHB, 5-nitroquinoxaline, and 6-mercaptopurine replication-dependent mutagenesis in *E. coli.*

b) Antimutagenesis against petite induction in *S. cerevisiae.*

c) Mitotic antimutagenic effect of the premeiotic DNA synthesis mutant spo-7 in *S. cerevisiae.*

d) Non-mutable mutants in *E. coli.*

e) Effects of IPTG on *laçZ* frameshift reversions in *E. coli.*

f) Antimutagenesis under anaerobic conditions.

g) Mutation-refractory strains of bacteria.

h) Antimutagenic effects of repair-defective mutants in *S. cerevisiae* and *N. crassa.*

Finally there are speculations regarding the effects of antimutagens on prophage induction, the occurrence of natural antimutagens in organisms producing known mutagens, and a possible preventative role of antimutagens in cancer induction.

Introduction

The topic of antimutagenesis in micro-organisms has been covered in a review completed approximately two years ago [2]. Thus there is now an opportunity to deal with additional results which have been published since then, and to mention findings which had been overlooked in writing the earlier review. One can also give, what seem now, on additional reflection, somewhat more appropriate interpretations of some antimutagenesis phenomena. There seems little point, however, in merely repeating the results and problems already dealt with in the 1975 article. Briefly the review covered the following topics of antimutagenesis against spontaneous and induced mutations: purine nucleosides (adenosine, guanosine, inosine), caffeine, manganous ions, methio-

Abbreviations: NHB = 4-nitro 6-hydroxybenzimidazole. 4 NQO = 4-nitroquinoline 1-oxide. IPTG = isopropyl-ß-thiogalactoside. UV = 254 nm ultraviolet light. EMS = ethyl methane sulphonate.

nine and histidine, spermine (and other polyamines), quinacrine and acridines, phenothiazine tranquilisers and dibenzocycloheptene antidepressants, coumadin (warfarin), actinomycin D, and basic fuchsin. The antimutagenic effects of chloramphenicol, 5-hydroxyuridine, 6-azauracil and pyronin-B on ultraviolet-induced mutational yields were also covered. Examples were quoted where genic repression or derepression (induction) had an antimutagenic outcome. At a more purely genetic level the review surveyed antimutagenic effects of the genetic background, and of repair-defective and DNA replication mutations in phage, bacteria and several eukaryotic microbes.

Antimutagenesis against NHB and other growth-dependent mutagens

One paper dealing with bacterial antimutagenesis which had been overlooked in writing the earlier review [2] was that of Greer [7], who studied mainly *pur*$^-$ → Pur$^+$ reversions and str^s → str^r forward mutations in *E. coli* strain W. Mutations induced by 4-nitro 6-hydroxybenzimidazole (NHB) during growth and replication, but not those arising spontaneously, were subject to antimutagenesis by the related compounds 4-hydroxy 6-nitrobenzimidazole and 4-hydroxy 6-nitrobenzotriazole. NHB was not mutagenic in non-growing bacteria. Furthermore, NHB induction of mutations during growth was counteracted by an antimutagenic effect of amino acid mixtures (e.g. serine + glutamine + methionine). Amino acid antimutagenesis also acted against the induction of mutations during growth by 5-nitroquinoxaline in this same *pur*$^-$ strain of *E. coli* W, but was not observed in another, *E. coli* B/r, strain. Perhaps not too surprisingly *met*$^-$ → Met$^+$ reversions induced, in growing cells of *E. coli* strain 15, by the base analogue 6-mercaptopurine were subject to an antimutagenic effect of added adenine, guanine, adenosine, or guanosine. A somewhat similar case, of suppression of the mutagenic activity of base analogues by added deoxyribonucleosides, was reported by Howard and Tessman [10] for phage S 13.

Mutation refractory strains of bacteria

A special case of the genetic control of antimutagenesis is that of mutation refractory organisms. For example *Micrococcus radiodurans* is non-mutable by UV, 4 NQO, γ-irradiation, mitomycin C or hydroxylamine, and only weakly mutable by nitrous acid, EMS or β-propiolactone. It is, however, mutable spontaneously and by nitrosoguanidine [31, 32]. Similarly *Proteus mirabilis* is not UV-mutable [17], and neither was one of the 22 strains of *Salmonella typhimurium* tested by Zinder and Lederberg [37]. There are, too, phages such as λ, T 1, T 3 and T 7 which are only mutable by UV when not only the phage but also the host cell is irradiated. These examples can probably all be understood in terms of the absence of error-prone repair systems capable of dealing with certain types of genetic damage [29, 35]. Rather, these mutagen-refractory strains possess solely error-free, and hence non-mutagenic, repair systems. Thus it would be of considerable interest if it proves possible to isolate mutable derivatives of normally non-mutable strains (the converse approach has, of course, been frequently adopted, e.g. *lex*, *recA*, *rad*, *umr* and *rev* mutants). Indeed the cases of those mutators which exhibit a synergistic effect with mutagens may be special examples of this, as too may be hypermutable strains. Eventually, however, the analysis of any such case of a change from an error-free to an error-prone type of repair will need to be made at a biochemical level, as has already been done for mutator, wild-type, and antimutator T 4 DNA polymerases [1, 6].

Miscellaneous antimutagenesis in *E. coli*

Using the approach pioneered by Lemontt in *S. cerevisiae* [15] there are now reports of the isolation, in *E. coli*, of mutants selected primarily on the basis of their reduced mutability (rather than having been isolated originally on the basis of repair — or recombination — deficiency). Cox [3] has reported mutants which can counteract the mutator effect of *mutT*, and there are some five loci (including two previously identified from repair-defective mutants) at which mutations can lead to decreased ultraviolet mutability (K. Brooks Low, personal communication).

At a physiological level, genic induction by IPTG has been found (Clarke and Shankel, this meeting) to reduce the frequency of both spontaneous and caffeine-induced reversions of *lacZ* frameshift mutant ND 160, arising during growth in *E. coli* K 12. This result parallels an earlier report of Herman and Dworkin [8] that one, exceptional, *lacZ* frameshift mutant reverted with ICR-191 approximately 7-fold less frequently in the presence of this inducer than when the lac operon was in the repressed state.

Anaerobic conditions

Novick [22] reported that growth of *E. coli* under anaerobic conditions is antimutagenic, against both spontaneous and theophylline-induced phage-resistant mutations arising in tryptophan-limited chemostat cultures. In view of these findings, which were ascribed at the time to a possible rise in intracellular concentrations of adenosine, it should be of interest to compare spontaneous and induced mutagenesis, in those bacteria and eukaryotic microbes which are facultative anaerobes, under anaerobic and anerobic conditions. A more speculative extension of this approach would be mutagenesis testing of bacterial mutants altered in their facultative anaerobic growth potential. Use could also be made in eukaryotic organisms such as *S. cerevisiae*, *Schizosaccharomyces pombe* and *Neurospora crassa* of cytoplasmic or nuclear respiratory-deficient mutants. To some, limited, extent this approach has already been attempted [23, 33] but not with the predicted antimutagenic outcome.

Antimutagenesis against "petite" induction.

A topic not covered in the earlier review [2] was that of antimutagenesis against cytoplasmic "petite" induction in *Saccharomyces cerevisiae*. There are several papers relating to this area. Nagai [19] showed that caffeine, at concentrations up to 1200 μg per ml, reduced the frequency of occurrence of spontaneous petite mutants (from approximately 49% down to approximately 4%) in populations of a highly unstable haploid strain, R6U7. Induction of *rho*$^-$ (cytoplasmic DNA-deficient) petites by 4NQO treatment of growing cells was found, in a later study by Nagai [20], to be largely prevented by pretreatment of the cells with eosin Y or related dyes. A paper by Moustacchi

[18] had shown that the induction of petites by 254 nm ultraviolet light was markedly reduced if the cells were pre-grown, before irradiation, in the presence of 1000 μg per ml concentrations of adenine, thymine and/or cytosine. Finally, the induction by ethidium bromide of *rho*⁻ mutants in non-growing cells of *S. cerevisiae* has been found, in two studies [9, 36] to be partially prevented by the presence of caffeine. There is some evidence that caffeine inhibits the degradation of an ethidium bromide-cytoplasmic DNA complex by a nuclease.

Antimutagenic effect of the *spo-7* mutant in *S. cerevisiae*

There has been reported, in *S. cerevisiae*, an intriguing example of a specific antimutagenic effect [5]. This involves the *spo-7* mutant which is blocked in premeiotic DNA synthesis. In a diploid strain homozygous for the *spo-7* mutation there is a 100-fold decrease in the frequency of mitotically-arising forward mutations at a tyrosine-inserting ochre suppressor (*t*RNA) locus. Such suppressor mutations lead to simultaneous suppression of the *lys2-1* and *ade2-1* markers. Interestingly there is no *spo-7* antimutagenic effect on spontaneous lys$^+$ or ade$^+$ intragenic back-mutations. (This study also gives an interesting example of a plating medium effect, in that Lys$^+$ Ade$^+$ phenotypic revertants, due to the ochre suppressor mutations, grow up on lysine-deficient but not on adenine-deficient medium [27]). It is not reported if *spo-7* acts antimutagenically against spontaneous, or indeed any types of induced, suppressor mutations arising in growing haploid cells.

Antimutagenic effects of repair gene mutants in *S. cerevisiae*

Since the earlier review there has become available a considerable amount of additional information regarding the involvement of repair genes in radiation and chemical mutagenesis in *S. cerevisiae*. This work has involved, mainly, a study of reversion of molecularly-defined *cyc-1* mutants [13, 14, 24, 25, 26, 30]. For example *cycl-131*, which reverts via GC → AT transitions, and *cycl-133* which reverts via GC → TA transversions were tested. The results obtained provided examples in the *rev3* and *rad6* loci which reduced UV revertibility of all the various test markers, i.e. generalised antimutagenic effect. However, in other cases any reduction in mutagenesis depended both on the particular *rad* allele and on the marker tested for revertibility, i.e. specific antimutagenic effect. It should be noted that all the comparisons of mutability between strains, even where these had very different sensitivities to the lethal effects of a given mutagen, were made at equal dose levels. In some, but by no means all, cases the conclusions reached would have been rather different if comparisons of mutability between strains had been made instead at equal survival levels. (However purely lethal effects can give a misleading impression of apparent antimutagenesis when mutational comparisons are made at identical survival levels). One presumes that comparisons of UV and γ mutability were also made at equal dose levels in the cases of the *N. crassa UVS* and other mutants quoted by SCHROEDER [28].

While in the review, and above, cases of genetic antimutagenesis have been considered to be those in which a mutant allele in a repair (or other) gene led to decreased mutability, exactly the opposite viewpoint is possible. Thus cases in which a repair mutant results in enhanced mutability, compared with the wild-type, often indicate that the locus in question codes, in the wild-type, for an antimutagenic, error-free, repair enzyme. Examples, for UV mutagenesis, are the *uvr A* locus in *E. coli*, the *rad* 1, 2, 3, 4 and 22 loci in *S. cerevisiae* (for *cycl-131* reversions), and the *UVS2* and *upr1* loci in *N. crassa*. Indeed several other cases of increased mutation rates can, perhaps, be viewed as resulting from a loss of an antimutagenic effect. Examples are the conditional mutator, *mutD5*, studied by DEGNEN and COX [4], and the mutator effect of a *dam* mutation in *E. coli* [16]. In the latter case one might regard the wild-type DNA methylation pattern as being antimutagenic in effect.

Speculations

Because of the parallels which exist between the processes of induced mutagenesis and prophage induction [29, 35], it would seem realistic to screen for antimutagens by testing for reduction or abolition of prophage induction. Conversely some already known antimutagens should decrease prophage induction. In the special case of pantoyl lactone, or cytidine plus guanosine, in a *tif* strain this is so [12, 34].

Secondly, a search for natural antimutagens (or highly effective error-free repair systems) might be instituted in these organisms known to contain, or to produce, natural mutagens. Examples are aflotoxins, mitomycin C, edeine, heliotrine, psoralens, methylated purines, or steroidal diamines.

Finally in the realm of cancer induction, if this should result from mutagenesis, or the insertion or release of viral genomes from host cell chromosomes, antimutagens might play a preventative role. To an extent this prediction has already been verified, in the case of caffeine diminution of 4NQO-induced cell transformation [11] and tumorigenesis [21].

Acknowledgments

I am most grateful to Professor Charlotte AUERBACH, F.R.S., Drs. Brian KILBEY and Pam McATHEY for interesting discussions and suggestion during a short visit to Edinburgh during August 1976.

References

[1] BESSMAN, M. J., N. MUZYCZKA, M. F. GOODMAN and R. L. SCHNAAR, Studies on the biochemical basis of spontaneous mutation. II. The incorporation of a base and its analogue into DNA by wild-type, mutator and antimutator DNA polymerases, J. Mol Biol., **88** (1974) 409–421.
[2] CLARKE, C. H. and D. M. SHANKEL, Antimutagenesis in microbial systems, Bact. Revs., **39** (1975) 33–53.
[3] COX, E. C., Mutator gene studies in *Escherichia coli*: the *mut T* gene, Genetics Suppl., **73** (1973) 67–80.
[4] DEGNEN, G. E. and E. C. COX, A conditional mutator gene in *Escherichia coli*: isolation, mapping and effector studies, J. Bacteriol., **117** (1974) 477–487.
[5] ESPOSITO, M. S., M. BOLOTIN-FUKUHARA and R. E. ESPOSITO, Antimutator activity during mitosis by a meiotic mutant of yeast, Molec. Gen. Genet., **139** (1975) 9–18.
[6] GOODMAN, M. F., M. J. BESSMAN and N. R. BACHUR, Adriamycin and daunorubicin inhibition of mutant T 4 DNA polymerases, Proc. Natl. Acad. Sci., **71** (1974) 1193–1196.
[7] GREER, S. B., Growth inhibitors and their antagonists as mutagens and antimutagens in *Escherichia coli*, J. gen. Microbiol., **18** (1958) 543–564.

[8] Herman, R. K. and N. B. Dworkin, Effect of gene induction on the rate of mutagenesis by ICR-191 in *Escherichia coli*, J. Bacteriol., **106** (1971) 543–550.
[9] Hixon, S. C. and K. L. Yielding, A protective effect of caffeine on the ethidium induced petite mutation in yeast, Mutation Res., **34** (1976) 195–200.
[10] Howard, B. D. and I. Tessman, Identification of the altered bases in mutated single-stranded DNA. II. In vivo mutagenesis by 5-bromodeoxyuridine and 2-aminopurine, J. Mol Biol., **9** (1964) 364–371.
[11] Kakunaga, T., Caffeine inhibits cell transformation by 4-nitroquinoline 1-oxide, Nature, **258** (1975) 248–250.
[12] Kirby, E. P., W. L. Ruff and D. A. Goldthwait, Cell division and prophage induction in *Escherichia coli*: effects of pantoyl lactone and various furan derivatives, J. Bacteriol., **111** (1972) 447–453.
[13] Lawrence, C. W. and R. Christensen, UV mutagenesis in radiation-sensitive strains of yeast, Genetics, **82** (1976) 207–232.
[14] Lawrence, C. W., J. W. Stewart, F. Sherman and R. Christensen, Specificity and frequency of ultraviolet-induced reversion of an iso-1-cytochrome c ochre mutant in radiation-sensitive strains of yeast, J. Mol. Biol., **85** (1974) 137–162.
[15] Lemontt, J. F., Genes controlling ultraviolet mutability in yeast, Genetics Suppl., **73** (1973) 153–159.
[16] Marinus, M. G. and N. R. Morris, Biological function for 6-methyladenine residues in the DNA of *Escherichia coli* K 12, J. Mol. Biol., **85** (1974) 309–322.
[17] Mates, M., Untersuchungen zur UV-induzierten Mutationsauslösung bei einigen Streptomycin-abhängigen Stämmen von *Proteus mirabilis* Hauser, Z. allg. Mikrobiol., **4** (1964) 22–41.
[18] Moustacchi, E., Effect antimutagène des bases puriques et pyrimidiques dans l'induction par les rayons ultraviolets de la mutation cytoplasmique "petite colonie" de la levure, C.R. Acad. Sci. Paris, **266** (1968) 720–723.
[19] Nagai, S., Stabilizing effect of caffeine on a respirationally unstable strain of yeast, Exper. Cell Res., **26** (1962) 253–259.
[20] Nagai, S., Counteracting effect of eosin and related dyestuffs on the production of respiration-deficient mutants in yeast by 4-nitroquinoline 1-oxide, Mutation Res., **34** (1976) 187–194.
[21] Nomura, T., Diminution of tumorigenesis initiated by 4-nitroquinoline 1-oxide by post-treatment with caffeine in mice, Nature, **260** (1976) 547–549.
[22] Novick, A., Mutagens and antimutagens, Brookhaven Symp. Biol., **8** (1956) 201–215.
[23] Paterson, H. F., Investigations into a reversal of diepoxybutane specificity in *Neurospora crassa*, Mutation Res., **25** (1974) 411–413.
[24] Prakash, L., Lack of chemically induced mutation in repair-deficient mutants of yeast, Genetics, **78** (1974) 1101–1118.
[25] Prakash, L., Effect of genes controlling radiation sensitivity on chemically induced mutations in *Saccharomyces cerevisiae*, Genetics, **83** (1976) 285–301.
[26] Prakash, L. and F. Sherman, Mutagenic specificity: reversion of iso-1-cytochrome c mutants of yeast, J. Mol. Biol., **79** (1973) 65–82.
[27] Queiroz, C., The effect of the plating medium on the recovery of nonsense suppressors in *Saccharomyces cerevisiae*, Biochem. Genet., **8** (1973) 85–99.
[28] Schroeder, A. L., Genetic control of radiation sensitivity and DNA repair in Neurospora, in Molecular Mechanisms for Repair of DNA, part B (P.C. Hanawalt and R. B. Setlow, editors), Plenum Publishing Corp., New York 1974, p. 567–576.
[29] Sedgwick, S. G., Inducible error-prone repair in *Escherichia coli*, Proc. Natl. Acad. Sci., **72** (1975) 2753–2757.
[30] Sherman, F. and J. W. Stewart, Variation of mutagenic action on nonsense mutants at different sites in the iso-1-cytochrome c gene of yeast, Genetics, **78** (1974) 93–113.
[31] Sweet, D. M. and B. E. B. Moseley, Accurate repair of ultraviolet-induced damage in *Micrococcus radiodurans*, Mutation Res., **23** (1974) 311–318.
[32] Sweet, D. M. and B. E. B. Moseley, The resistance of *Micrococcus radiodurans* to killing and mutation by agents which damage DNA, Mutation Res., **34** (1976) 175–186.

[33] Von Borstel, R. C., S. K. Quah, C. M. Steinberg, F. Flury and D. J. C. Gottlieb, Mutants of yeast with enhanced spontaneous mutation rates, Genetics Suppl., **73** (1973) 141–151.
[34] Witkin, E. M., Thermal enhancement of ultraviolet mutability in a *tif-1 uvrA* derivative of *Escherichia coli* B/r: evidence that ultraviolet mutagenesis depends upon an inducible function, Proc. Natl. Acad. Sci., **71** (1974) 1930–1934.
[35] Witkin, E. M., Persistence and decay of thermoinducible error-prone repair activity in non-filamentous derivatives of *tif-1 Escherichia coli* B/r: the timing of some critical events in ultraviolet mutagenesis, Mol. Gen. Genet., **142** (1975) 87–103.
[36] Wolf, K. and F. Kaudewitz, Effect of caffeine on the rho$^-$ induction with ethidium bromide in *Saccharomyces cerevisiae*, Molec. Gen. Genet., **146** (1976) 89–93.
[37] Zinder, N. D. and J. Lederberg, Genetic exchange in Salmonella, J. Bacteriol., **64** (1952) 679–699.

Symposium II

Dose-Effect Relationships, Treshold Effects, and the Question of Chronic Versus Acute Exposure

Chairman's Introduction

B. A. Bridges

M. R. C. Cell Mutation Unit, University of Sussex, Falmer,
Brighton BN 1 9 QG, England

Of all the areas of environmental mutagenesis, that concerned with the response to low doses of mutagens, particularly as it relates to human hazards, is at once the most difficult and the most controversial. Mutations are stochastic events and whereas each one is an "all or nothing" event and any individual genome may escape damage, in a large population there is always a finite probability of a mutation even at very low mutation rates. As Dr. Šram points out, this is in confrontation with the toxicologists common experience of thresholds.

Of course, the situation is more complex than a simple confrontation. The law of mass action tells us that for agents covalently binding to or in other ways reacting chemically with DNA, there cannot in principle be a threshold. The slope of the dose response may, however, be so altered by secondary processes that a quasi-threshold response results. Such secondary processes may occur after the reaction with DNA, for example through DNA repair processes, or before it, for example through detoxification processes (see Bridges and Stamper, Mutation Res., 33 (1975) 87—91). Evidence for either of these is almost non-existent in man.

On the other hand, there are mutagens (for example, caffeine and 8-methoxypsoralen in the dark) which do not react covalently with DNA but interact in a non-covalent manner to increase the probability of certain errors during DNA metabolism. There is no reason why the kinetics of such an effect should not have a threshold.

I hope that the different viewpoints in this field will find expression during the Symposium. I would merely caution against becoming committed to dogma in the absence of evidence, and urge the search for common ground. The history of science shows clearly that controversy flourishes in the absence of good data; and science without good data is bad science.

Comparison of Replication Errors and Mutagenic Repair as a Source of Spontaneous Mutations

R. C. von BORSTEL[1], P. J. HASTINGS[1] and Cornelia SCHROEDER

[1]Department of Genetics, University of Alberta
Edmonton, Alberta T6G 2E9, Canada and
Institut für medizinische und allgemeine
Mikrobiologie, Virologie und Epidemiologie
des Medizinischen Bereichs der Humboldt-Universität zu Berlin
Lehrstuhl Virologie
DDR 104 Berlin, Schumannstr. 20/21

Introduction

It has been suggested that spontaneous mutations are a product of DNA metabolism resulting from replication, repair, and meiotic processes. Meiosis is implicated because higher mutation rates occur during meiosis than mitosis (MAGNI and VON BORSTEL, 1962) and there is a correlation of recombinant chromosomes during meiosis with spontaneous mutational events (MAGNI, 1963). Replication is implicated because some mutations of DNA polymerase of bacteriophage T 4 have enhanced spontaneous mutation rates (SPEYER, KARAM and LENNY, 1966) and some have depressed spontaneous mutation rates (DRAKE and ALLEN, 1968). Repair is implicated because some radiation-sensitive mutants have enhanced spontaneous mutation rates (BÖHME, 1967). Those mutants which have enhanced spontaneous mutation rates are called mutators, and these have been thought to be the key to understanding the origin of spontaneous mutations.

The channelling hypothesis for explaining mutator activity in *Saccharomyces cerevisiae*

The concept of the channelling of spontaneous lesions into mutagenic repair pathways (HASTINGS, QUAH and VON BORSTEL, in press; VON BORSTEL and HASTINGS, in press) was devised to explain the mutator activity of numerous mutator loci in *Saccharomyces cerevisiae*. This hypothesis states that spontaneous lesions in DNA may be repaired by any one of a number of metabolic routes which have evolved to repair damaged DNA. If a repair pathway is blocked genetically, then the lesions may become the substrate for another repair pathway (unless the lesion is irreparable by any other pathway, in which case death of the cell would take place). If lesions are channelled into a mutagenic (error-prone) repair pathway, then the spontaneous mutation rate would be enhanced. Thus a mutator can be a genetic block at the beginning of a pathway, or at a branch point, where the substrate for repair is diverted into a mutagenic repair pathway.

[1] Research supported by the National Research Council of Canada and grant 21 294 from the National Institute of General Medical Sciences.

The concept of channelling offers an explanation for four major phenomena: (1) why there is a large number of mutant loci with mutator activity in yeast, (2) why nearly all mutator mutants in yeast are associated with repair, (3) why, when the first step of an error-free radiation repair pathway is blocked genetically, this first-step mutant is a mutator, and (4) why mutator mutants are recessive.

The channelling hypothesis requires that there be more than one mutagenic repair pathway in order to explain why *rad6*, the gene encoding the putative first step in the mutagnic repair pathway, has mutator activity, and why nitrous acid and nitrosoimidazoladone induce mutations in *rad6* strains (PRAKASH, 1974). Moreover, the channelling hypothesis offers a possible explanation as to why respiration-deficient strains of yeast have mutator activity: respiration-deficient strains of yeast do not exhibit liquid-holding recovery (LYMAN and HAYNES, 1967), and this blocked repair would be expected to channel spontaneous lesions into one of the mutagenic repair pathways (VON BORSTEL and HASTINGS, in press). Further, the channelling hypothesis explains a puzzling phenomenon that has been seen in experiments carried out over a fifteen-year period. On the basis of the doubling of DNA in a diploid strain, one might expect the spontaneous mutation rate in diploid strains to be double that of haploid strains. Indeed, this doubling of rate has been observed (MAGNI and VON BORSTEL, 1962; GOTTLIEB and VON BORSTEL, 1976). In other experiments the spontaneous mutation rates in diploid strains have been essentially the same as (VON BORSTEL, CAIN and STEINBERG, 1971), or even lower than (QUAH, unpublished data), that of the parental haploid strains. It has been known for a number of years that diploid strains heterozygous for the mating type alleles, a and α, are more resistant to γ-radiation than strains homozygous a/a and α/α (MORTIMER, 1958; LASKOWSKI, 1960). This may be explained by postulating a repair system which has come to be known as a/α repair. One would thus expect a/α diploids to have a lower spontaneous mutation rate than haploids since they have available to them a mode of repair (presumably non-mutagenic) which is unavailable in haploid cells. Thus a lower proportion of spontaneous lesions will be processed by mutagenic repair.

None of the above phenomena can be explained easily by a model for mutator activity which involves a defective DNA polymerase complex.

Mutator activity in bacteriophage T4 caused by defective DNA polymerase

A great advance toward explaining spontaneous mutations at the molecular level in bacteriophage T 4 took place when BRUTLAG and KORNBERG (1972) discovered the 3'-exonuclease editing activity of the DNA polymerase encoded by gene 43. MUZYCKA et al. (1972) presented evidence that enhanced spontaneous mutation rates can be associated with a defect in 3'-exonuclease activity and that a depressed spontaneous mutation rate can be associated with a defect in the DNA polymerase activity itself.

The lack of the editing function of 3'-exonuclease in bacteriophage T 4 as a prerequisite for mutator activity cannot be the whole answer even in bacteriophage T 4. HERSHFIELD and NOSSAL (1973) showed that one mutator mutant of DNA polymerase had a higher exonuclease-polymerase ratio than the wildtype enzyme, a property reserved for antimutator DNA polymerases by SCHNAAR et al. (1973). In this case, "discrimination against wrong nucleotides occurs primarily during incorporation" (HERSHFIELD and NOSSAL, 1973). It is certain that a mutated DNA polymerase could be disturbed in the relation between its incorporating and editing function as BESSMAN and his collaborators (1974) attest, and it seems possible that a mutation of DNA polymerase in bacteriophage T 4 which disturbs discrimination could be a source of transversions. RIPLEY (1975) has demonstrated that mutated bacteriophage T 4 DNA polymerase can indeed be a source of transversions.

Bacteriophage T 4 has both non-mutagenic and mutagenic repair systems (GREEN and DRAKE, 1974), albeit not particularly powerful ones. Since bacteriophage T 4 has a number of mutator genes besides gene 43, then channelling of spontaneous lesions into mutagenic repair could possibly account for some of the spontaneous mutations in this organism also (DRAKE 1973).

The situation in *Escherichia coli*

The characteristics of mutators which we take as evidence that much spontaneous mutation arises by mutagenic repair in yeast can also be seen in *E. coli.*

E. coli has mutagenic repair, as seen by the dependence of UV mutagenesis on the *recA* gene product (WITKIN, 1969; MIURA and TOMIZAWA, 1968). It is known that proflavin-induced mutagenesis proceeds by mutagenic repair pathways (KOHNO and ROTH, 1974) in *Salmonella.* There is some evidence that the same is true for base-analog mutagenesis in *E. coli* as well (PIETRZYKOWSKA, 1973; PIETRZYKOWSKA, LEWANDOWSKA and SHUGAR, 1975), although the first base-analog experiments of Pietrzykowska are disputed by WITKIN and PARISI (1974). It would not appear to be necessary, *a priori,* for an analog to take on an enol, imino or syn form (TOPAL and FRESCO, 1976) for it to be recognized as a lesion by a repair system.

Mutations of *polA*$^-$ show mutator activity (COUKELL and YANOFSKY, 1970; VACCARO and SIEGEL, 1975; KONDO, 1973), but it is not easy to suppose that this is in any way analogous to mutators in gene 43 of T 4. For the following reasons it seems more likely that mutagenic repair is involved: (a) The *polA*$^-$ alleles are UV sensitive, (b) All seven alleles of *polA* for which the spontaneous mutation level has been reported are mutators, (c) The mutator activity predominantly causes frameshift and deletion, not substitution, mutations (COUKELL and YANOFSKY, 1970; VACCARO and SIEGEL, 1975; KONDO, 1973), (d) The *polA*$^-$ alleles interact with several other mutators. There is reported to be a synergism in mutator activity seen with *mutL, mutR* and *mutS* (COX, 1976). Such synergism can be interpreted, by analogy with UV survival synergism in yeast (GAME and COX, 1973), to mean that two different repair pathways, which compete for the same substrate, have been closed, and that the repair processes which remain are mutagenic. The double mutant *polA*$^-$ *mutU*$^-$ is not viable, implying that there is an essential function which may be performed by either of two systems.

These properties of *polA*$^-$ may be correlated by supposing that *polA*$^-$ is involved in gap filling in a variety of situations, in both normal cell growth and in repair. If any of these gaps are left, they are a substrate for mutagenic repair.

The *mutT*⁻ gene is recessive, and is active only in replication (Cox, 1970), so that *mutT*⁺ is most easily understood at present as an error-reducing component of the polymerase complex (Cox, 1976).

E. coli appears to differ significantly from yeast in that yeast has little or no involvement of repair systems in normal vegetative growth, since no lethal combinations of *rad* or *mut* alleles have been reported. It is worth noting, however, that several *rad* strains of yeast are meiotic lethals (see CATCHESIDE, 1974).

Conclusion

Evidence gained from the properties of different mutator loci indicate that channelling of spontaneous lesions into mutagenic repair pathways can account for most of the spontaneous mutations in *Saccharomyces cerevisiae*, and a substantial portion of those in *Escherichia coli*. In a microorganism where a mutagenic repair pathway does not exist, as may be the case for *Proteus mirabilis* (BÖHME, unpublished data), mutagenic repair cannot be a source of spontaneous mutations. The loss of editing function or alteration of discrimination in DNA polymerase appears to account for much of the mutator activity in bacteriophage T4, as demonstrated by the identification of antimutators in gene 43 mutants, at least until other antimutators are found.

It may be that proportionately more of the mutators found in higher organisms concern repair pathways because there are more repair pathways.

The relative importance of mutagenic repair as a source of spontaneous mutations in yeast and *E. coli* must await the identification of antimutator activity in these systems.

References

BESSMAN, M. J., N. MUZYCKA, M. F. GOODMAN and R. L. SCHNAAR, Studies on the biochemical basis of spontaneous mutation. II. The incorporation of a base and its analogue into DNA by wild-type, mutator and antimutator DNA polymerase, J. Mol. Biol., **88** (1974) 409–421.

BÖHME, H., Genetic instability of an ultraviolet sensitive mutant of *Proteus mirabilis*, Biochem. Biophys. Res. Commun., **28** (1967) 191–196.

BRUTLAG, D. and A. KORNBERG, Enzymatic synthesis of deoxyribonucleic acid. XXXVI. A proofreading function for the 3'→5' exonuclease activity in deoxyribonucleic acid polymerases, J. Biol. Chem., **247**, (1972) 241–248.

CATCHESIDE, D. G., Fungal genetics, Ann. Rev. Genetics **8** (1974) 279–300.

COUKELL, M. B. and C. YANOFSKY, Increased frequency of deletions in DNA polymerase mutants of *Escherichia coli*, Nature **228** (1970) 633–635.

COX, E. C. Mutator gene action and the replication of bacteriophage λ DNA. J. Mol. Biol., 50 (1970) 129–135.

COX, E. C. Bacterial mutator genes and the control of spontaneous mutation. Ann. Rev. Genetics, **10** (1976) 135–156.

DRAKE, J. W. and E. F. ALLEN, Antimutagenic DNA polymerases of bacteriophage T 4, Cold Spring Harbor Symp. Quant. Biol., **33** (1968) 339–344.

DRAKE, J. W., The genetic control of spontaneous and induced mutation rates in bacteriophage T 4, Genetics, **73** (Suppl.) (1973) 45–64.

GAME, J. C. and B. S. COX, Synergistic interaction between *RAD* mutations in yeast, Mutation Res., **20** (1973) 35–44.

GOTTLIEB, D. J. C. and R. C. VON BORSTEL, Mutators in *Saccharomyces cerevisiae*: *mut 1–1*, *mut 1–2* and *mut 2–1*, Genetics, **83** (1976) 655–666.

GREEN, R. R. and J. W. DRAKE, Misrepair mutagenesis in bacteriophage T 4, Genetics, **78** (1974) 81–89.

HASTINGS, P. J., S.-K. QUAH and R. C. VON BORSTEL, 1976, Spontaneous mutation by mutagenic repair of spontaneous lesions in DNA, Nature (in press).
HERSHFIELD, M. S. and N. G. NOSSAL, *In vitro* characterization of a mutator T4 DNA polymerase, Genetics, (Suppl.) **73** (1973) 131–136.
KOHNO, T. and J. R. ROTH, Proflavin mutagenesis of bacteria, J. Mol. Biol., **89** (1974) 17–32.
KONDO, S., Evidence that mutations are induced by errors in repair and replication, Genetics, (Suppl.) **73** (1973) 109–122.
LASKOWSKI, W., Inaktivierungsversuche mit homozygoten Hefestämmen verschiedenen Ploidiegrades. I. Aufbau homozygoter Stämme und Dosiseffektkurven für ionisierende Strahlen, UV und organische Peroxyde, Z. Naturforsch., **15b** (1960) 495–506.
LYMAN, J. T. and R. H. HAYNES, Recovery of yeast after exposure to densely ionizing radiation, Radiation Res. Supplement, **7** (1967) 222–230.
MAGNI, G. E., The origin of spontaneous mutations during meiosis, Proc. Natl. Acad. Sci. (USA), **50** (1963) 975–980.
MAGNI, G. E. and R. C. VON BORSTEL, Different rates of spontaneous mutations during mitosis and meiosis in yeast, Genetics, **47** (1962) 1097–1108.
MIURA, A. and J. TOMIZAWA, Studies on radiation-sensitive mutants of *E. coli*. III. Participation of the Rec system in induction of mutation by ultraviolet irradiation, Molec. Gen. Genetics, **103** (1968) 1–10.
MORTIMER, R. K., Radiobiological and genetic studies on a polyploid series (haploid to hexaploid) of *Saccharomyces cerevisiae*, Radiation Res., **9** (1958) 312–326.
MUZYCKA, N., R. L. POLAND and M. J. BESSMAN, Studies on the biochemical basis of spontaneous mutation. I. A comparison of the deoxyribonucleic acid polymerases of mutator, antimutator and wildtype strains of bacteriophage T4, J. Biol. Chem., **247** (1972) 7116–7122.
PIETRZYKOWSKA, I., On the mechanism of bromouracil-induced mutagenesis, Mutat. Res., **19** (1973) 1–9.
PIETRZYKOWSKA, I., K. LEWANDOWSKA and D. SHUGAR, Liquid-holding recovery of bromouracil-induced lesions in DNA of *Escherichia coli* CR-34 and its possible relation to dark-repair mechanisms, Mutat. Res., **30** (1975) 21–32.
PRAKASH, L., Lack of chemically induced mutation in repair-deficient mutants of yeast, Genetics, **78** (1974) 1101–1118.
RIPLEY, L. S., Transversion mutagenesis in bacteriophage T4, Mol. Gen. Genet., **141** (1975) 23–40.
SCHNAAR, R. L., N. MUZYCKA and M. J. BESSMAN, Utilization of aminopurine deoxynucleoside triphosphate by mutator, antimutator and wildtype DNA polymerases of bacteriophage T4, Genetics (Suppl.), **73** (1973) 137–140.
SPEYER, J. F., J. D. KARAM and A. B. LENNY, On the role of DNA polymerase in base selection, Cold Spring Harbor Symp. Quant. Biol., **31** (1966) 693–697.
TOPAL, M. D. and J. R. FRESCO, Complementary base pairing and the origin of substitution mutations. Nature, **263** (1976) 285–289.
VACCARO, K. K. and E. C. SIEGEL, Increased spontaneous reversion of certain frameshift mutations in DNA polymerase I deficient strains of *Escherichia coli*, Molec. Gen. Genetics, **141** (1975) 251–262.
VON BORSTEL, R. C., K. T. CAIN and C. M. STEINBERG, Inheritance of spontaneous mutability in yeast, Genetics, **69** (1971) 17–27.
VON BORSTEL, R. C. and P. J. HASTINGS, Mutagenic repair pathways in yeast, Proc. VII Intern. Cong. Photobiol. (Rome, Italy, August 29-September 3, 1976) (in press).
WITKIN, E. M., The mutability toward ultraviolet light of recombination-deficient strains of *Escherichia coli*, Mutat. Res., **8** (1969) 9–14.
WITKIN, E. M. and E. C. PARISI, Bromouracil mutagenesis: Mispairing or misrepair?, Mutat. Res., **25** (1974) 407–409.

Gene Mutations: Dose-Response Relationships and their Significance for Extrapolation to Man

S. Hussain and L. Ehrenberg

University of Stockholm
Department of Radiobiology
Wallenberg Laboratory
Lilla Frescati
S-10405 Stockholm 50
Sweden

Introduction

Different pro- and eukaryotic systems are in use to test the genetic toxicity of environmental chemicals and there is a constant search going on for improved systems which are more relevant to man. This is so, because the problem we are facing today is to make some assessment of the compounds in question as to their ability to cause genetic damage to the present or future human generations on the basis of data obtained from these test systems. The aim of the present discussion is to show that comparatively easily obtainable chemico-biological information could be tentatively used for this purpose, as a ground-work for later improvements. To do this the chemical dose has to be defined.

Dose

In order to know the magnitude of an effect as a function of the amount of the agent in question it is imperative to have some measure of this agent expressed as dose. Whereas for ionizing radiations this is easily expressed in terms of the energy absorbed per unit weight of the tissue, the dose of the mutagenic or the cancerogenic chemical has to be defined so that it could be useful for risk estimates. Of the different expressions for chemical dose (cf. Ehrenberg et al. 1974), we are interested in the one termed as the tissue dose.

The tissue dose (D) is defined as the time integral of a chemical or its reactive metabolites in the target tissue, and is expressed as follows:

$$D = \int_t [RX]dt \tag{1}$$

(D = dose: RX = the active species, [ ] denoting concentration in M; t = time).

Statement of the dose in this way allows for the fact that the chemicals vary with respect to rates of absorption, distribution, metabolism and excretion.

The amount of the reaction products of the agent, RX, taking part in the reaction below, can be experimentally determined.

$$RX + Y^- \rightarrow RY + X^- \tag{2}$$

(RX = agent; Y^- the attacking nucleophile; RY = the product or the alkylated species).

From a knowledge of, e.g., the amount of alkylated cysteine or histidine per gram of some protein, in mammals for instance hemoglobin, or the amount of 7-alkylguanine per g of DNA, it is possible to estimate the dose, as defined above (OSTERMAN-GOLKAR et al. 1976; EHRENBERG and OSTERMAN-GOLKAR 1976), according to

$$D = \frac{[RY]}{g(Hb\ or\ DNA) \cdot k_y}$$

where k_y is the rate constant for the respective reaction.

Dose-response curves

Information on the shape of dose-response curves is basic to any risk evaluation. These curves obtained with most of the agents are actually indicated to contain linear components in the low dose region. Previously, the barley was used to study the mutagenicity of chemicals by EHRENBERG et al. (1966), but the system was not sensitive to low doses of the agents used. Therefore, the need arose to study the mutation curves with bacterial systems.

TURTÓCZKY and EHRENBERG (1969) used the *E. coli* Sd-4 and observed linearity in the low dose region for mutation to non-dependence (see e.g., Figs. 1 and 2 of the quoted paper for ethyl methanesulfonate and isopropyl methanesulfonate). In a follow-up study, different agents were tested and a linear relationship was observed in the low dose treatment with methyl methanesulfonate, ethyl methanesulfonate, ethylene oxide, diepoxybutane, chloroacetone, chloroacetaldehyde, chloroethanol, triethylene melamine and nitrogen mustard.

HESLOT (1962) tested a number of agents on the yeast, *Schizosaccharomyces pombe*. All the epoxides, most of the sulfonates and the sultones yielded reverse mutation curves with linear components in the low dose region.

The demonstration of a linear dose response for mutation makes it meaningful to use the tissue dose as a basis in estimates of genetic risk in the practically interesting range of low doses of mutagenic pollutants.

Knowing the rate constant for the reaction of an alkylating agent, *RX*, with some standard nucleophile (such as water or cysteine), and knowing the substrate given as $n \approx 2$. This value is not significantly different from that of Gua-0[6] but may also represent a mean value of several group of low reactivity in the DNA.

Correction factor for functionality

The finding that the mutation frequency is proportional to the degree of alkylation is tentatively taken to be valid for simple alkylating agents. In special cases correction factors have to be developed for a quantitative assessment of the risk. This is indicated to be necessary for steric factors, especially the intercalating agents, and for chemical reactivity of introduced alkyl groups (*R* in eqn. 1), as in the case of polyfunctionality. The latter problem will be discussed here.

We had been faced with the problem as to whether vinyl chloride could act as a bifunctional agent. The evidence so far accumulated in our laboratory showed that its effective metabolite is chloroethylene oxide and not chloroacetaldehyde. Alkylated products of chloroethylene oxide contain an aldehyde group (GÖTHE et al. 1974) which might enter into further reaction, e.g. with other groups in the DNA.

In order to be useful in risk estimates, a test system must respond to functionality of different kinds. It might be expected that back-mutation systems responding, e.g., to base-pair transition, would be less able than a forward mutation system to indicate risk from further reaction of alkyl groups introduced. In agreement with this expectation such systems, for example the commonly used *Salmonella typhimurium* strains, were found to be less sensitive to resolve the differences in functionality, as seen for ethylene-oxide, diepoxybutane and triethylene melamine (Table 1) representing mono-, bi- and

Table 1

Comparison of the response of *Salmonella typhimurium* and *E. coli* systems to agents differing with respect to functionality.

Mutagenic agent	Bacterial strain	Mutation frequency $\times 10^8$ per mM. h
Ethylene oxide	*S. typhimurium* G 46	1
	S. typhimurium TA 1535	21
	E. coli Sd-4	2
Diepoxybutane	G 46	1
	TA 1535	27
	Sd-4	32
Triethylene melamine	G 46	17
	TA 1535	1500
	Sd-4	10000

trifunctional agents respectively. Whereas *Salmonella typhimurium* G-46 and TA 1535 bring out no difference, diepoxybutane is 16 times more effective than ethylene oxide in *E. coli* Sd-4. In the same way the effectiveness of triethylene melamine is still more magnified in the latter system. Similarly, it was reported that on the basis of *Salmonella* system the mutagenic effectiveness of ethylene oxide and chloroethylene oxide was the same per alkylation (RANNUG et al. 1976). In contrast to this the latter agent was about 20 times more effective in the *E. coli* system. This means that due attention should be given to the reactivity of the aldehyde group.

It should be pointed out that in the barley chlorophyll mutation system which measures forward mutation in several hundred genes, the influence of functionality is observable, too. Thus the difunctional diepoxybutane has been reported to be some 50 times more effective than the monofunctional ethylene oxide (EHRENBERG and GUSTAFSSON 1957). In the back-mutation system in *Neurospora* no difference in the effectiveness of the two agents was found (KÖLMARK and WESTERGAARD 1953).

It is known that in general bi- or trifunctional agents are much more effective than monofunctional ones with respect to the inhibition of tumoral growth (ROSS 1950), induction of cancer (VAN DUUREN 1969), mutation (EHRENBERG and GUSTAFSSON 1957) and the induction of chromosomal aberrations (MOUTSCHEN-DAHMEN and LOPPES 1963).

A test system to be used for quantitative risk estimates has to respond correctly to the consequences of functionality irrespective of its mechanism (cross-linking etc.). Using ethylene oxide as a standard monofunctional agent, the relative effectiveness of diepoxybutane and triethylene melamine per alkylation was calculated (Table 2). The

Table 2

Mutans per unit alkylation of DNA $\left(\frac{\text{Mutation frequency}}{10^{-3}\text{M.h} \times k_{n=2}}\right)$, $\times 10^8$

	Ethylene oxide	Diepoxybutane	Triethylene melamine
E. coli Sd-4	143	2300 (×16)	360000 (×2500)
S. typhimurium G-46	72	72 (×1)	620 (×9)
TA 1535	1500	1900 (×1.3)	55000 (×37)
Barley[1]	0.03%[1]	1% (×~30)[1]	20%[1] (×3000)

[1] Final frequencies per mM. h; experiment done at 20°. Correction factors given in terms of relative effect within brackets.

correction factors, obtained by taking the quotient of mutation frequencies per alkylation due to the compared agents, are given in this Table. It is interesting to note that for triethylene melamine the *E. coli* and barley systems yield the same, high correction factor for trifunctionality. Also in the *Salmonella* systems triethylene melamine shows a relative effectiveness per alkylation >1. Due to difficulties of estimating the reaction rates, of this compound partly because of pH dependence, the figures given for triethylene melamine have to be taken as approximative. The main thing demonstrated is the fact that triethylene melamine is two orders of magnitude more effective in the forward mutation systems than in the back-mutation systems.

It should be observed that there is a correction factor 16 for bifunctionality on the basis of the *E. coli* system, and it was mentioned that chloroethylene oxide and ethylene oxide differed by a factor of about 20. This would imply that vinyl chloride, or strictly its metabolite chloroethylene oxide could be acting as a bifunctional agent.

Risk evaluation and extrapolation to man

It should be pointed out that there is a difference between qualitative risk estimates and quantitative risk assessments. The *Salmonella* system developed by Ames is for the qualitative purpose and as such serves well. Since we are interested in the quantitative assessment of risk, we look for some system which could yield the data for this purpose, hence we have tentatively used the *E. coli* system combined with the chemical analysis. Though the *E. coli* Sd-4 system has been observed to respond to a wide range of agents and behaves like a forward system the need for some reliable eukaryotic forward mutation system cannot be overemphasized.

The degree of alkylation of some centers with low nucleophilic strength determined by chemical analysis (Ehrenberg et al. 1974; Osterman-Golkar et al. 1976) has been related to the frequencies of mutations in bacteria and a counterpart of the radiation dose yielding the same amount of genetic damage determined. In this way a translation factor has been obtained which may be expressed as follows:

$$1 \text{ rad-equivalent} \leftrightarrow \frac{[RY]}{[Y]} = 10^{-7} \quad \text{at n} = 2 \tag{4}$$

The same translation factor was found to be valid for barley and an examination of data on dominant lethals, translocations and specific locus mutations in mice (EHLING 1973; MOUTSCHEN 1969; Committee 17, 1975), revealed that it is approximately valid also in the mouse (cf. OSTERMAN-GOLKAR 1975). It seems therefore justified at least as a first approximation, to use the relationship (4) for risk estimates in man. Values of the tissue dose required for the calculation of $\frac{[RY]}{[Y]}$ at n = 2 might be obtained from the degree of alkylation of cellular macromolecules such as hemoglobin (OSTERMAN-GOLKAR et al. 1976).

Applying the relationship (4) for a risk assessment would hence give

$$\text{Genetic risk} = \frac{[RY_{n=2}[}{[Y_{n=2}]} \times 10^7 \times f_1 \times f_2$$

where f_1 is a correction factor for functionality according to Table 2 of this paper. The factor f_2 is introduced to allow for variations in dose between organs, e.g., the ratio of gonal dose to blood dose. For many small low-molecular-weight alkylating agents $f_2 \approx 1$.

Acknowledgement

The work was supported by the Swedish Atomic Research Council.

References

[1] Committe 17 appointed by the Council of Environmental Mutagen Society, Science, **187** (1975) 503.
[2] EHLING, U. H., 1st Int. Conf. on Environ. Mutagens, Asilomaar 1973.
[3] EHRENBERG, L. and Å. GUSTAFSSON, Hereditas, **43** (1957) 593.
[4] EHRENBERG, L., U. LUNDQUIST, S. OSTERMAN and B. SPARRMAN, Hereditas, **56** (1966) 277.
[5] EHRENBERG, L., K. D. HIESCHE, S. OSTERMAN-GOLKAR and I. WENNBERG, Mutat. Res., **24** (1974) 83.
[6] EHRENBERG, L. and S. OSTERMAN-GOLKAR, 1st Europ. Symp. on Rad-equivalence, Inst. du Radium, Orsay, 1976 (EURATOM. in press).
[7] GÖTHE, R., C. J. CALLEMAN, L. EHRENBERG and C. A. WACHTMEISTER, Ambio, **3** (1974) 234.
[8] HESLOT, H., Abhandl. Deut. Akad. Wiss. Berlin Kl. Med., (1962) 193.
[9] HUSSAIN, S. and L. EHRENBERG, Int. J. Radiat. Biol., **27** (1975) 355.
[10] HUSSAIN, S., L. EHRENBERG and G. AHNSTRÖM, Hereditas, **83** (1976) 134.
[11] KÖLMARK, G. and M. WESTERGAARD, Hereditas, **39** (1953) 209.
[12] MOUTSCHEN-DAHMEN, J., M. MOUTSCHEN-DAHMEN and R. LOPPES, Nature, **199** (1963) 406.
[13] MOUTSCHEN, J., Mutat. Res., **8** (1969) 581.
[14] OSTERMAN-GOLKAR, S., Doctoral Thesis, Stockholm 1975.
[15] OSTERMAN-GOLKAR, S., L. EHRENBERG and C. A. WACHTMEISTER, Radiat. Botany, **10** (1970) 303.

[16] Osterman-Golkar, S., L. Ehrenberg, D. Segerbäck and I. Hällström, Mutat. Res., **34** (1976) 1.
[17] Rannug, U., R. Göthe and C. A. Wachtmeister, Chem.-Biol. Interactions, **12** (1976) 265.
[18] Ross, W. C. J., J. Chem. Soc., (1950) 2257.
[19] Turtóczky, I. and L. Ehrenberg, Mutat. Res., **8** (1969) 229.
[20] van Duuren, B. L., Ann. N. Y. Acad. Sci., **163** (1969) 633.

Genetic Aspects of Dose-Effect-Relationships and their Significance for the Extrapolation to Man-Chromosome Aberrations

G. Röhrborn and A. Basler

Department of Human Genetics and Anthropology;
University of Düsseldorf

1) Approximately 50 out of 10000 human newborns have chromosome aberrations. 40 of these have new chromosome and genome mutations. The frequency of embryos with chromosome aberrations may be estimated higher, because numerical aberrations are suspected to cause early lethality in most cases.

2) This assumption is confirmed by the high frequency — more than 35% — of chromosomal aberrations in clinical abortions.

3) Therefore it can be concluded, that at least 7% and at maximum 40% of all human zygotes have chromosome aberrations..

Chromosome and genome mutations in germ cells are very dangerous because they manifest themselves in the first generation and their consequences are severe.

Besides chromosome aberrations occuring in germ cells also chromosome aberrations in somatic cells may have serious consequences for man. For years mutagenesis and cancerogenesis have been correlated. Mutated somatic cells are suspected to be the first step in cancer development.

In chronic myelogenous leukaemia, cells have been found bearing a chromosome with a small deletion, the so-called "Philadelphia"-chromosome [10, 11, 18]. Furthermore, in a variety of tumours certain types of chromosome aberrations have been observed [4]. Monoclonal tumour development has been demonstrated for myoma of the uterus [8] and for lymphoma, myeloma, chronic granulocytic leukaemia, multicentric liver and colon carcinoma [3, 6]. The autosomal recessive diseases Bloom's syndrome, Fanconi's anaemia and ataxia teleangiectasia are associated with a high frequency of chromosome breakage and reunion, i.e., chromosome mutation. Moreover, patients suffering from these diseases run an increased risk of developing leukaemia and several types of cancer [15, 8, 16]. Mibelli's porokeratosis is an autosomal dominant trait associated with malignant tumours (squamous cell carcinomas). The patients lymphocytes have normal chromosomes. Fibroblasts cultured from biopsies of areas with skin lesions, however, showed increased frequencies of chromosome gaps, breaks and reunions, whereas biopsies of non-affected skin usually did not show such chromosomal aberrations [17, 7].

For these reasons mutagenicity testing of chromosome aberrations should be done in mammalian test systems, in somatic as well as in germ cells.

Possibly organotropic effects are the reason to perform cytogenetic tests, especially in those tissues which are entered by the compound. If there is no possibility to perform such a distribution study, various other tissues should be investigated in order to avoid false negative results.

One of the most important parameter in toxicology and mutagenicity is the dose effect relationship and the question of threshold effects. Test systems, intendent to analyse chromosome aberrations, should satisfy the main conditions of toxicology:

1) In the test system some compounds should not be active at all even in high doses. Here for instance caffeine should be mentioned. Neither in bone marrow [12] nor in spermatocytes (1) induced chromosome aberrations could be found by giving only caffeine to mammals in extremely high doses.

2) It should be possible to investigate dose effect relationships in the test system if toxic or mutagenic activity is to be recognized.

In test systems which indicate the frequency of induced chromosome aberrations in mammals directly (by chromosome analysis) or indirectly (by micronucleus and dominant lethal assay) those dose effect relationships could be demonstrated by an increasing number of authors. A few authors like Holden et al. (1973) even postulated thresholds for inducing chromosome aberrations in mammalian systems by 6-mercaptopurin. Some investigations performed in our institute will be discussed before we come back to these problems of thresholds:

In a comparing study we tested Cyclophosphamide, Trophosphamide and Isophosphamide in the bone marrows of Chinese hamsters (*Cricetulus griseus*) after various intervals following the last of 2 i.p. applications. The both injections were given 24 h apart in physiological solution. The dose given was 3/10 of the DL 50, because the toxic dose-effect-curve showed a similar slope for the three substances (Brock, pers. communication). It can be seen (Fig. 1) that the maximum effect is achieved 12 h after the last application. There was a slight decrease in the 24 h interval and a strong decrease in the 48 h interval.

% aberrant metaphases in bone marrow of Chinese hamster

interval	isophosphamide %	trophosphamide %	cyclophosphamide %	Control %
12 h	67 (74)	63.5 (68)	55 (63.5)	
24 h	41.5 (47)	31.9 (44)	35.5 (42.5)	1.0 (2.5)
48 h	16 (22.5)	12.5 (16.25)	8.5 (13.5)	

dose: 3/10 of DL_{50} (I = 180, T = 60, C = 120 mg/kg)
number of animals per group: 8
number of metaphases per animal: 200
with parenthesis = inclusive gaps
without parenthesis = exclusive gaps

Fig. 1. % of aberrant metaphases in bone marrow of Chinese hamster after treatment with isophosphamide, trophosphamide and cyclophosphamide.

Subsequent to this study our co-worker MOHR investigated the chromosomes of the bone marrow of Chinese hamsters after i.p. treatment with 2 injections, 24 h apart. The animals were sacrificed either 12 or 24 h after the last application, respectively. A dose-effect-curve can be seen in the 12 h interval (Fig. 2). However, with respect to the single types of aberrations there was no dose response for gaps and a very weak one for breaks. The correlation was best pronounced for exchanges. This demonstrates that sometimes dose effect relationships may be expressed only in some types of chromosome aberrations.

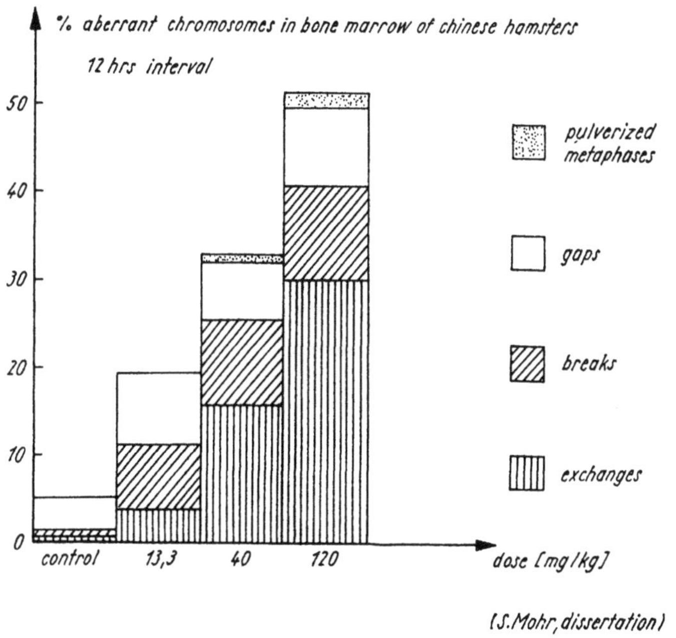

Fig. 2. % aberrant metaphases in bone marrow of Chinese hamsters 12 hrs after the last of two cyclophosphamide-injections.

Fig. 3 shows the results 24 hrs after the last application with cyclophosphamide. Again there was no correlation between dose and gaps. Furthermore the dose-response for breaks and exchange figures could only be demonstrated in the mean dose level. In the highest dose there was observed a negative dose effect, that means a decrease of aberrant metaphases. This might be a cytotoxic effect which results in a bone marrow depression. The micronucleus test, performed with the same animals confirmed this assumption. The frequencies of polychromatic erythrocytes in relation to the normochromatic ones were extremely decreased in the highest dose, especially 24 hrs after the last application. It should be mentioned that in these experiments the micronucleus test is directly comparable to the chromosome analysis, because from each animal one femur was used for chromosome analysis and the other one for the micronucleus test.

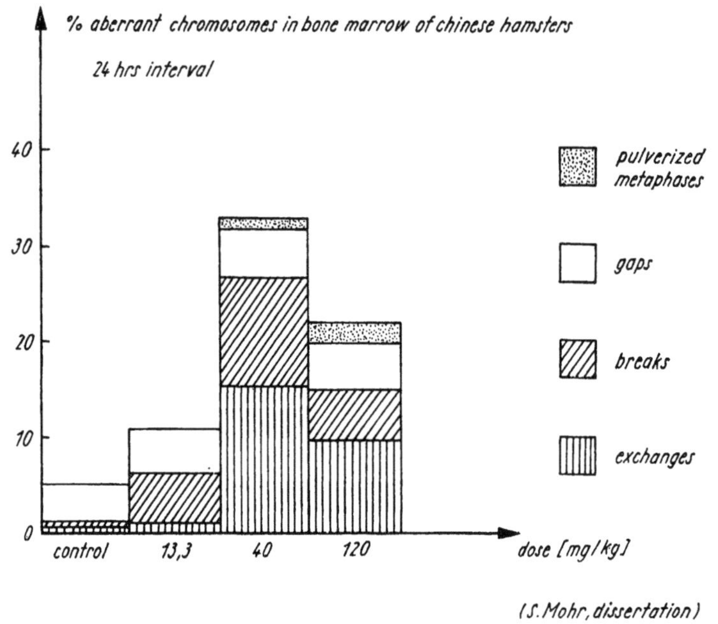

Fig. 3. % aberrant metaphases in bone marrow of Chinese hamsters 24 hrs after the last of two cyclophosphamide-injections.

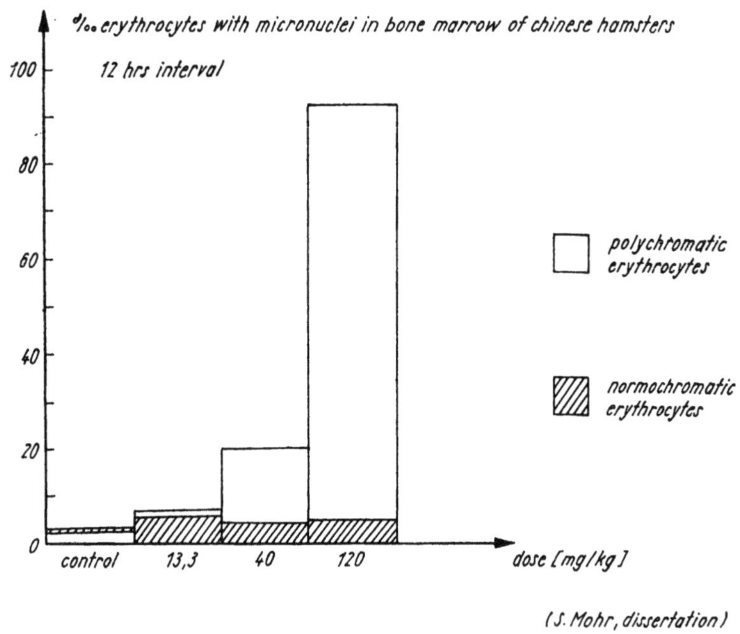

Fig. 4. ‰ erythrocytes with micronuclei in bone marrow of Chinese hamsters 12 hrs after the last of two cyclophosphamide-injections.

As can be seen (Fig. 4) the dose-effect-curve of micronuclei of polychromatic erythrocytes corresponds to the cytogenetic results. There is a dose effect relationship with respect to polychromatic micronuclei in the 12 h interval. A correlation with respect to micronuclei in polychromatic erythrocytes is also to be seen 24 hrs after the last application (Fig. 5).

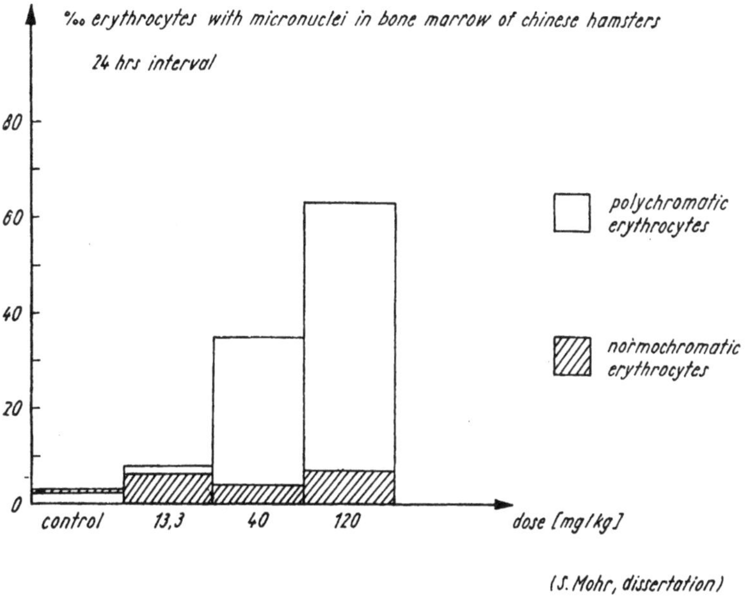

Fig. 5. ‰ erythrocytes with micronuclei in bone marrow of Chinese hamsters 24 hrs after the last of two cyclophosphamide-injections.

These dose-effect-curves, however, can only be found in Chinese hamsters. In different species, one should be aware that there might be differences. For example we analysed the mutagenic activity of Cyclophosphamide, Isophosphamide and Trophosphamide in mice, rats and Chinese hamsters. The results are to be seen in figure 6—8. The animals were sacrificed 6 hrs after the last application in all cases.

In mice (Fig. 6) and rats (Fig. 7) there was no dose-response after application of Isophosphamide and Trophosphamide. Only in Chinese hamsters (Fig. 8) a dose-effect-curve for Cyclophosphamide as well as for Isophosphamide and Trophosphamide is to be seen even 6 hrs after the last application. These differences may be due to different activation time for the three compounds in the three species tested.

In another experiment we analysed the chromosomes of bone marrows of Chinese hamsters after treatment with 3,4 benzo(a)pyrene (2).

The hamsters were administered orally with 2 × 450 mg/kg B(a)P, (experiment a + b), 5 × 180 mg/kg B(a)P (experiment c) 5 × 450 mg/kg (experiment d). It can be seen (Fig. 9, 10) that the frequency of aberrant metaphases exclusive gaps is increased significantly only when the hamsters have been administered with the highest dose of 5 × 450 mg/kg. Including gaps this could already be seen in experiment c.

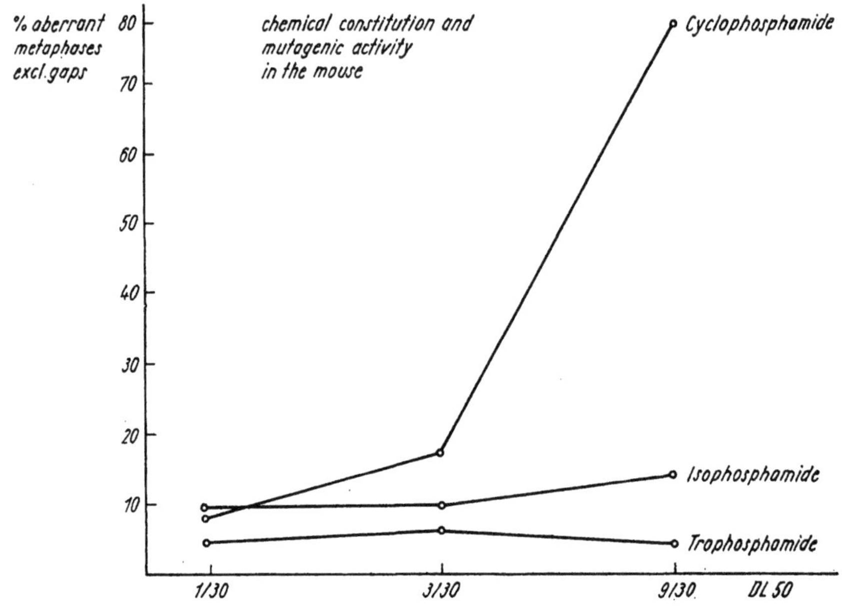

Fig. 6. Mutagenic activity of cyclophosphamide, isophosphamide and trophosphamide in bone marrow of mice 6 hrs after the last application.

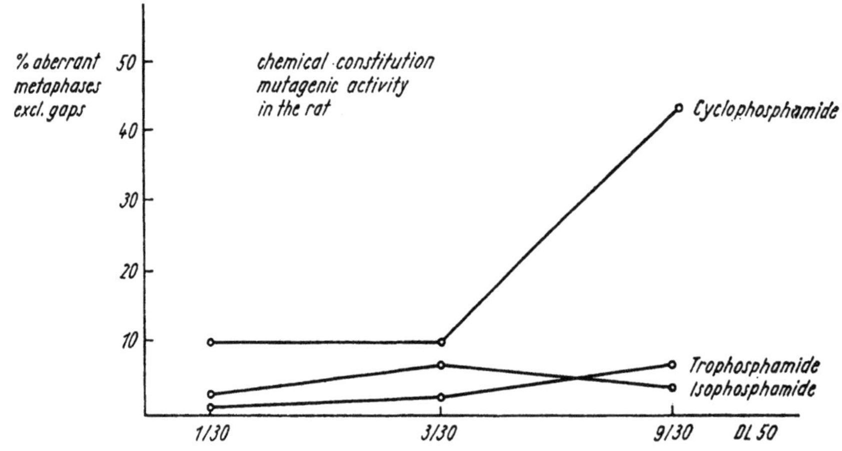

Fig. 7. Mutagenic activity of cyclophosphamide, isophosphamide and trophosphamide in bone marrow of rats 6 hrs after the last application.

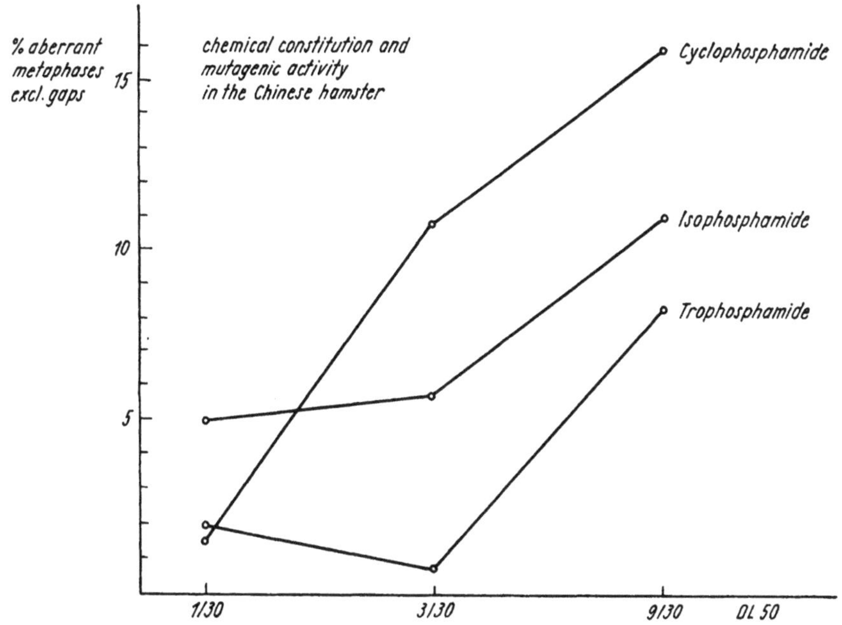

Fig. 8. Mutagenic activity of cyclophosphamide, isophosphamide and trophosphamide in bone marrow of Chinese hamsters 6 hrs after the last application.

Chromosome aberrations of Chinese hamster bone marrow cells
a and b) 2 × application of 450 mg/kg B(a)P
a) preparation 6 hrs after the last B(a)P application
b) preparation 24 hrs after the last B(a)P application
c) 5× application of 180 mg/kg B(a)P
d) 5× application of 450 mg/kg B(a)P

series	n metaphases analysed	aberrant incl. gaps n	%	p	aberrant excl. gaps n	%	p	n metaphases with breaks a. fragments	exchanges
control pool	3,200	34	1.06		5	0.15		5	0
control	600	7	1.16		2	0.33		2	0
experiment a	600	12	2.00		5	0.83		3	2
experiment b	800	20	2.50	1	4	0.50		3	1
experiment c	400	24	6.00	1 2 3	2	0.50		2	0
experiment d	1600	205	12.81	1 2 3 4	51	3.18	1 2 3 4	46	5

[1] $p \leq 0.02$ in comparision to the control pool
[2] $p \leq 0.02$ in comparision to the control
[3] $p \leq 0.02$ in comparision to the experiment a and b
[4] $p \leq 0.02$ in comparision to the experiment c

Fig. 9. Mutagenic activity of 3,4 benzo(a)pyrene in bone marrow of Chinese hamsters.

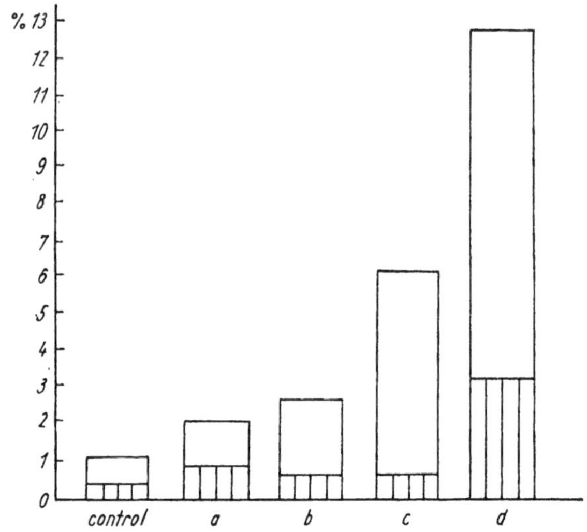

Structural chromosome aberrations of Chinese hamster bone marrow cells including and excluding ▯▯▯▯ gaps

Fig. 10. Mutagenic activity of 3,4 benzo(a)pyrene in bone marrow of Chinese hamsters.

Comparing these results we can state:
2 applications of cyclophosphamide result in a constant dose-effect-curve in bone marrow chromosomes of Chinese hamsters. With respect to 3,4 benzo(a)pyrene no dose-effect is to be recognized after application of low doses. In the highest dose, however, there is to be seen a sudden increase of aberrations, that means a threshold.

For benzene in the same test system we did not find any mutagenic effect, neither in the chromosome analyses nor in the micronucleus test. The only effect we found was a toxic one:

1) In the highest dose of 1 ml/kg 50% of the animals died.

2) The rate of polychromatic erythrocytes to normochromatic ones decreased with increasing doses (Fig. 11). In the lowest dose (0.25 ml/kg) the proportion of polychromatic to normochromatic erythrocytes is 6000 to 2962. In the highest dose, however, the relation is only 6000:843.

Micronuclei of Chinese hamster after benzene treatment

dose ml/kg	ratio of normochromatic/ polychromatic erythrocytes	‰ polychromatic erythrocytes with micronuclei	‰ normochromatic erythrocytes with micronuclei
0.25	6000/2962	1.8	0.7
0.5	6000/2883	0.8	1.3
1	6000/843	2	1.1
control	5533/5168	1.5	1

Fig. 11. Ratio of normochromatic/polychromatic erythrocytes and ‰ of erythrocytes with micronuclei in bone marrow of Chinese hamsters after benzene treatment.

It would be overhaste to conclude that the threshold dose for benzene is too high to detect chromosome aberrations at all, as to be seen in the next experiment. Female NMRI-mice were pretreated with pregnant mares'serum and with human chorionic gonadotropin (HCG) to synchronize the oestrus. Three hrs, after the injection of HCG 0,5 ml/kg benzene was applicated i.p. Chromosome analysis was done 16 hrs after the HCG-application.

Fig. 12 shows that benzene did not induce non-disjunction processes. The number of aneuploid metaphases was about the same as in the control. On the other hand there was a drastic increase of metaphases with chromosome aberrations inclusive gaps and exclusive gaps. This demonstrates that — if at all — a threshold could be claimed only for a certain test-system under certain conditions.

a) Frequencies of aneuploid metaphase II oocytes

strain	series	% aneu-ploidies	% m II with chromatids <40	>40
NMRI	control	8.0	6.0	2.0
NMRI	experiment	8.8	5.5	3.3

b) Types and frequencies of structural aberrant metaphase II oocytes

strain	series	n m II	% structural aberrant m II incl. gaps	exl. gaps	n metaphases with: gaps	breaks and fragments	centric fusions	interchanges
NMRI	control	122	0.82	0.82	0	0	1	0
NMRI	experiment	120	17.50	12.50	6	5	3	7

Fig. 12. Mutagenic activity of benzene in unfertilized NMRI-mice oocytes.

All these experiments show the following results:

1) Some compounds, like Cyclophosphamide, lead to a clear dose-response in some test-systems and animals, like bone marrow cells of Chinese hamsters.

2) Sometimes dose-effect relationships may only be expressed in some type of chromosome aberrations.

3) Some compounds result in no dose effects for some species, whereas it is to be seen for other species after treatment with the same compound.

4) Sometimes a clear threshold can be recognized.

5) In different tissues of one species threshold might be different.

The experiments lead to the general considerations on threshold-effects: In general toxicology it is a fact, that a threshold-dose exists frequently and there is no toxic effect below. This is quite understandable, because a low dose may damage permanently or temporarily very few cells and this damage can be compensated by other cells or by cell divisions. For these reasons we suggest to use the term "tolerable dose" instead of "threshold-dose".

On the contrary, with respect to induced chromosome aberrations and point mutations, we have to consider the following possibility: Only one of 10000 germ cells might have an induced mutation. If this specific germ cell comes to fertilization, each cell of the embryo will contain the same type of mutation. Subsequently the sickness of this person can be referred to the induced mutation. At least, with respect to chromosome aberrations, such a slight mutation rate can never be excluded by any statistical method.

Further difficulties are enhenced by the fact that cytogenetic studies in human in vivo are performed in general with lymphocyte cultures after exposition to mutagens. It has been demonstrated by ENGELHARD (5) in a comparative cytogenetic study in Chinese hamsters that lymphocyte cultures are at least less sensitive than bone marrow cells.

Nevertheless it has been shown that doses being effective in animals are also active in man. In our study for instance, 14 mg/kg Cyclophosphamide induced chromosome aberrations in bone marrow of Chinese hamsters. In human daily doses of 15 to 30 mg/kg were found to induce chromosome aberrations in lymphocytes of cyclophosphamide treated patients as well [14].

With respect to extrapolation to man, we cannot avoid to use dose-response-data in mammals for such extrapolations. We should be aware, however, that, as in other fields of toxicology, this extrapolation is very sophisticated and needs a safety factor. This safety factor should be higher the more people are exposed to the mutagen.

References

[1] ADLER, I.-D. and G. RÖHRBORN, Investigation of meiotic chromosomes of male mice after chronic caffeine treatment, Humangenetik, **8** (1969) 81–85.

[2] BASLER, A. and G. RÖHRBORN, Chromosome aberrations in oocytes of NMRI-mice and bone marrow cells of Chinese hamster induced with 3,4 benzopyrene, Mut. Res., **38** (1976) 321–332.

[3] BEUTLER, E. et al., Value of genetic variants of glucose-6-phosphate dehydrogenase in tracing the origin of malignant tumors, New. Engl. J. Me., **276** (1967), 389–391.

[4] DE GROUCHY, J. and DE NAVA, C., A chromosomal theory of carcinogenesis, Ann. intern. Med., **69** (1968) 381–389.

[5] ENGELHARD, G., Vergleichende Untersuchungen zur Sensibilität von Knochenmark, Lymphozyten und Spermatogonien des Chinesischen Hamsters gegenüber Butyl-Nitroso-Harnstoff. Dissertation, 1975.

[6] FIALKOW, P. S. et al., Clonal origin of chronic myelocytic leukemia in man, Proc. nat. Acad. Sci. (Wash.), **58** (1967) 1468.

[7] HARNDEN, D. G., Ataxia-telangiectasia Mibelli. Dissertation, Heidelberg 1973.

[8] HECHT, F., Leukaemia and lymphocytes in ataxia-telangiectasia, Lancet, **2** (1966) 1193.

[9] LINDER, D. and S. M. GARTLER, Glucose-6-phosphate dehydrogenase mosaicism: utilization as cell marker in study of leiomyomas, Science, **150** (1965) 67–69.

[10] NOWELL, P. C. and D. A. HUNGERFORD, Chromosome studies on normal and leukemic human leucocytes, J. nat. Cancer Inst., **25** (1960a) 85–109.

[11] NOWELL, P. C. and D. A. HUNGERFORD, A minute chromosome in human granulocytic leukemia, Science, **132** (1960b) 1497.

[12] RÖHRBORN, G., Correlation between mutagenesis and carcinogenesis. In: Transplacental carcinogenesis, Lyon, International Agency for Research on Cancer, pp. 168–174 (1973) (IARC Scientific Publications No. 4).

[13] RÖHRBORN, G. and U. BUCKEL, Investigations on the frequency of chromosome aberrations in bone marrow cells of chinese hamsters after simultaneous application of caffeine and cyclophosphamide, Hum. Genet., **33** (1976) 113–119.

[14] Schmid, E. and M. Bauchinger, Comparison of the chromosome damage induced by radiation and cytoxan therapy in lymphocytes of patients with gynaecological tumours, Mut. Res., **21** (1973) 271–274.
[15] Schroeder, T. M. et al., Spontane Chromosomenaberrationen bei familiärer Panmyelopathie, Hum. Genet., **1** (1964) 194–196.
[16] Schroeder, T. M., Genetische Faktoren der Krebsentstehung, Fortschr. Med., **16** (1972) 603–608.
[17] Taylor, A. M. R. et al., Chromosomal instability associated with suspectibility to malignant disease in patients with porokeratosis of Mibelli, J. nat. Cancer Inst., **51** (1973) 371–378.
[18] Tijo, J. H. et al., The Philadelphia chromosome and chronic myelogenous leukemia, J. nat. Cancer Inst., **36** (1966) 567–584.

Dose-Response Relationship: Viewpoints of Toxicology

Zdeněk Bardoděj

Medical Faculty of Hygiene, Charles University, Praha
Czechoslovakia

I. Threshold

Paracelsus definition [21] formulated in 1537, that "all substances are poisons, the dose being a cause that makes a substance loose its toxicity" had been far from being doubted through four centuries. This rather pessimistically tuned conclusion became later modified and used more frequently in the sense that the dose is a cause of toxic effect-"dosis facit venenum". This very old thruth still remains, as a matter of fact, a basic principle, employed both in pharmacology and environmental toxicology, because if it were in principle wrong, then all the hygienic standards, including the MAC values established for working as well as living environment, would become unthinkable.

It was only after the Second World War when Druckrey [9, 10] has come to a conclusion, that Paracelsian definition fails to hold true for all the effects and that there are chemical substances inducing even in very low doses irreversible changes, which gradualy accumulate, without any detectable effect at first, but with serious consequences occuring after a certain period of latency; the latency period tends to become reduced with increasing total dose of administered substance. The well known equation derived by Druckrey

$$D \cdot T^n = K,$$

where D = dose, T = time and K = constant for a given effect was accepted by a WHO —Scientific group in its report on the Assessment of the carcinogenicity and mutagenicity of chemicals, that was released, in 1974 [4]. The exponent n of this equation was found to be higher than one in all carcinogenic compounds tested. Here we can hardly speak about a true threshold dose or a no-effect dose. According to the Committee 17 [6] appointed by the Council of the U.S. Environmental Mutagen Society no information was available in 1975 proving or disproving the existence of threshold response to a mutagen in any system.

It is very difficult, and at present nearly impossible, to recommend any experimentally or epidemiologically provable safe limits for mutagens and carcinogens: we can only accept their doses or levels in the environment with calculated or predictable risks. This practice has been used also in the evaluation of ionizing radiation hazards [1, 20].

The threshold for environmental mutagens and carcinogens have not been unequivocaly proved, however, there are some indications that it might exist for instance thanks to the DNA-repair processes. It has been also observed, that the risk of pulmonary cancer is reduced in ex-smokers, which supports the idea that these effects may be partly reversible [4].

The scepticism to the attempts to establish non-effect levels led the participants of the Second International Symposium on Maximum Allowable Concentrations, held in Paris in 1963, to a conclusion that no MAC values for carcinogens could be recommended [19]. However, acceptable this approach might seem from the scientific point of view it cannot get along with practical aspects. It is well known that the proof of absence of a substance in the environment depends on the sensitivity of the method used for its detection or determination. It is why the lists of MAC values keep still limits for mutagens and carcinogens. So, MAC for chlorinated biphenyl, "Aroclor" was recommended for international use at the Joint ILO/WHO Committee on Occupational Health meeting in Geneva in 1968 [13] and the limit for vinyl chloride was not excluded from the list in spite of its known carcinogenicity and mutagenicity.

In Czechoslovakia, where the regulations on chemical hazards are very strict, about one quarter of 102 substances with established MAC values has been at least once in the literature claimed to have some carcinogenic and/or mutagenic potencial (acrolein, ammonia, arsenic, benzene, beryllium, cadmium oxide, carbon tetrachloride, chlorinated biphenyl, chloroform, chloroprene, chromium, cobalt, cyclohexylamine, dicyclohexylamine, nitrite, formaldehyde, hydrazine, lead, maleic anhydrine, mercury, ozone, phenol, toluidine, trichlorethylene).

Some problems linked with the evaluation of genetic and carcinogenic hazards are the same as in the case of other toxic substances, e.g. mechanism and kinetics of absorption, distribution, transport, biotransformation, and elimination. Mutagenic and carcinogenic activities are only particular cases of the biological activity of a substance. That's why we speak about genetic toxicology. Differences can be detected in the mode of action, in the irreversibility or low reversibility of induced changes, in delayed effects, in the methods and test systems used for the detection of these activities, in the extent of late consequences, frequently even after exposures to extremely low doses, and in the present scarcity of our knowledge as to how to extrapolate laboratory data to human population.

II. Dose-response relationship

The dose-effect and dose-response relationships are the most important data in every toxic activity in practice. The dose-effect relationship, i.e. the "graded effect", is the relationship between exposure and the severity of qualitatively specified effect in an individual and is very important mainly for pharmacologists and clinical toxicologists. The dose-response relationship, i.e. a quantal response, is the relationship between exposure and the relative number of individuals with a quantitatively specified severity of a qualitatively specified effect in a group of subjects. The quantal response is more important for hygienists, environmental toxicologists, but mainly for those dealing with mutagenesis and carcinogenesis. The confidence with which a mutagenic or carcinogenic effect can be established is strenghtened by the observation of an increasing incidence with increasing exposure.

Two types of dose-response relationship are possible in toxicology: A straight-line relationship, where any quantity of the substance, even a single molecule, will induce an effect; no threshold exists in this case (Fig. 1). A sigmoid relationship, where no response at all is detectable at sufficiently low doses, provided that a certain threshold value is not reached (Fig. 2). Physiologists and toxicologists are in general inclined to accept this sigmoid-type dose-response relationship also for mutagens and carcinogens. DINMAN [8] suggested in 1972, when considering the number of molecules required

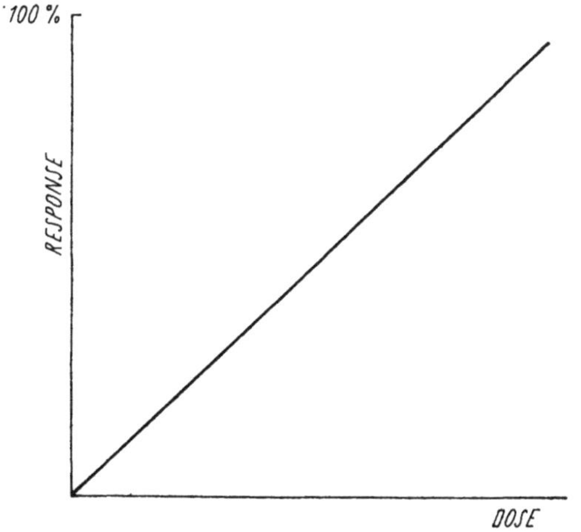

Fig. 1. A straight-line dose-response relationship: no threshold exists.

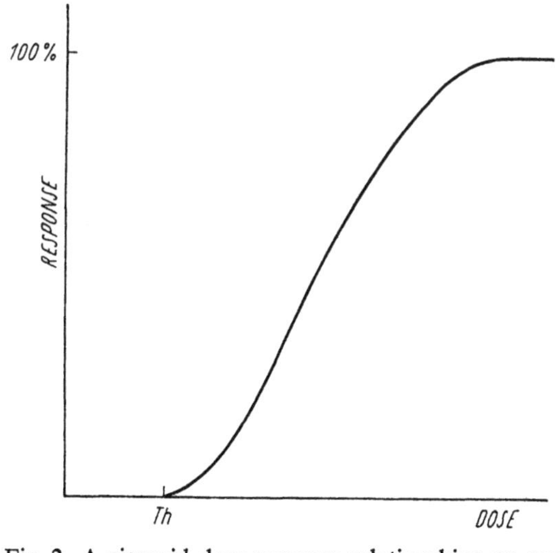

Fig. 2. A sigmoid dose-response relationship: no response below a threshold (Th).

to enter a cell before one of them will react with a specific site, that there must theoretically exist a dose, below which the carcinogen is ineffective. FREESE [12] in 1973, when analyzing mutagenic response, remarks that neither for radiation nor for carcinogens that react directly with DNA has the dose dependence indicated the presence of a threshold and comments, that for different reasons an apparent or true threshold of mutation or cancer induction can be expected to occur, e.g. when the active compound is inactivated before it could reach nucleus, or in a weak mutagen, that inhibits some enzyme needed to produce a precursor of DNA. It can be mentioned here that the mutagenic effect of a substance appears to be higher after a single dose than after fractionated doses of the same mutagen dose administered during a prolonged period. CHANDLER [5] in 1974 developed a chemical-kinetic kinetic model of the dose-response relationship in mutagenesis and carcinogenesis based on the fact, that mutations are the results of a sequence of chemical reactions, which must be studied by methods of chemical kinetics. CROW [7] and BRIDGES [4] tried to link chemical mutagenesis with radiation and suggested to assess chemical mutagens in terms of roentgen equivalent dose (RED) or in rad-equivalents (radeq). The mutagenicity of three compounds was calculated in REC units (rem equivalent-chemical) by the Committee 17 [6]: EHRENBERG and co-workers [11] expressed the mutagenic activity of ethylene oxide in rad-equivalents; the calculated genetic risk of persons exposed to ethylene oxide was claimed to be higher than that occuring at the maximum permissible dose of radiation, i.e. 0.1 rad/week.

In toxicology we often use the coefficients of safety to minimize the potential risk in very sensitive individuals. The steeper the dose-response relationship, the lower the safety margin needed to provide equivalent assurance of safety. In carcinogenic and mutagenic effects this is not possible for the reason mentioned above and it is also preferrable to replace the term of threshold dose or zero-effect dose by acceptable or tolerable dose or — permissible dose or — allowable dose, which means basically the same without implying that a drug is ineffective beyond a certain dose limit.

The statistical methodological problem how to extrapolate animal data when establishing acceptable doses of vinyl chloride for humans was very carefully analyzed by SCHNEIDERMAN, MANTEL and BROWN [15]. Their conclusions were rather pessimistic.

III. Research and practice

Another very important problem to be shortly mentioned is concerned with how to introduce the results of laboratory research and epidemiological studies into the practice. New or strong mutagens and carcinogens should be avoided completely, but what to do with old and weak carcinogens and mutagens which are important for our way of life? The exposure to them has to be minimized by changes in technology and in drugs by careful examination of risk-benefit relationship.

In a general effort to ensure healthy environment there has been elaborated and published quite a number of systems for the evaluation and control of chemicals to be industrially used, e.g. by Committee 17 [6], Prof. BRIDGES [4], Dr. SCHÖNEICH [16], and Dr. BORA [3]; a unified system of mutagenicity testing has been recommended in the USSR and Czechoslovakia [2]. There is a good chance, that hazardous food additives and pesticides will be eliminated in the near future.

The present situation in industrial plants is, however, different. Tons of chemicals whose mutagenicity has been already proved or can be suspected, are and will be produced and used. Bis-chloroethylethers, alkyl sulfates, epichlorohydrin, chloroamines and ethylene oxide can be mentioned here; other may result from an accident or may be formed by a unpredictable synthesis. There are, indeed, available excellent and valuable screening tests in microorganisms that are not time consuming as well as tests on mammals But I am affraid that quite a number of physicians rather mistrust animal data at all preferring only observations in humans. Species-specific susceptibility to chemical substances is already well known. So it is not surprising, that H. F. SMYTH [18], who in 1959 evaluated different methods used for establishing TLV, found short-term inhalation or injection experiments in animals the worst; but animal data after repeated inhalation or feeding were as good as human experience in industry.

In the first years after the Second World War the occupational exposure of workers producing the same goods in different plants was very different. With the progress in technology no such great differences have been usually found since fifties between different factories even in various countries. The advancement in technology and the existence of hygienic standards are primary factors reducing the extent of occupational exposure — mainly in big industrial plants. In Czechoslovakia, we have conducted routine follow-up studies in workers occupationally exposed to lead, benzene, ethyl benzene, toluene, styrene, and trichloroethylene, analysing air and urine samples, for two decades at least. The exposures have tended to decline. Some of these data are useful also today. As trichloroethylene was reported to be carcinogenic [17] a study is now being in progress analysing exposure to trichloroethylene and cancer incidence in dry cleaning workers [14].

The research in mutagenesis and carcinogenesis develops surprisingly fast. All the new discoveries in the field of genetic toxicology should be now efficiently and without unjustifiable delay introduced into practice to ensure protection of the people involved. This is, however, far from being easy. The effect e.g. of chlorine inhalation manifests itself immediately and the relationship between exposure and acute symptoms of poisoning cannot be overlooked; the carcinogenicity of vinyl chloride was discovered after thirty years of its use and there are plants, where the atmospheric concentrations are relatively high five years after this discovery.

Much more papers dealing with genetic toxicology and published not only in Mutation Research, but also in periodicals for general practitioners, pediatricians, industrial hygienists, pharmacologists and chemists are urquently needed. The relationships between risk-benefit and cost-benefit are difficult to evaluate, if the severity of the problem is not recognized.

Not only managers in industry, but also some physicians underestimate the degree of risk, although some of them see the hand writing on the wall.

References

[1] Advisory committee on the biological effects of ionizing radiations, The effects on populations of exposure to low levels of ionizing radiations. National Academy of Sciences-National Research Council, Washington, D.C., 1972.

[2] Bochkov, N. P., R. J. Šrám, N. P. Kuleshov and V. S. Zhurkov, System for the evaluation of the risk from chemical mutagens for man: basic principles and practical recommendations, Mutation Res. (in press).
[3] Bora, K. C., Modified tier approach to mutagenicity testing, International symposium on new developments in mutagenicity testing of environmental chemicals, Žinkovy Castle, 1970.
[4] Bridges, B. A., Some general principles of mutagenicity screening and a possible framwork for testing procedures. Environ. Health Perspectives, No. 6 (1973) 221.
[5] Chandler, J. Lr., The chemical dynamics of mutagenesis and carcinogenesis. I. A chemical-kinetic kinetic model of dose-response relationships, Mutation Res., **25** (1974) 169–178.
[6] Committee 17, Environmental mutagenic hazards. Mutagenicity screening is now both feasible and necessary for chemicals entering the environment, Science, **187** (1975) 503–514.
[7] Crow, J. F., Impact of various types of genetic damage and risk assessment, Environ. Health Perspectives, No. 6 (1973) 1–5.
[8] Dinman, B. D., "Non concept" of "no threshold". Chemicals in the environment, Science, **175** (1972) 495.
[9] Druckrey, H., K. Küpfmüller, Dosis und Wirkung, Editio Cantor, Aulendorf i. Wttbg., 1949.
[10] Druckrey, H., Grundlagen der toxikologischen Methodik, Arzneimittel-Forsch., **7** (1957) 449–456.
[11] Ehrenberg, L., K. D. Hiesche, S. Osterman-Golkar and I. Wennberg, Evaluation of genetic risk of alkylating agents: tissue doses in the mouse from air contaminated with ethylene oxide, Mutation Res., **24** (1974) 83–105.
[12] Freese, E., Thresholds in toxic, teratogenic, mutagenic, and carcinogenic effects, Environ. Health Perspectives, No. **6** (1973) 171–176.
[13] Joint ILO/WHO committee on occupational health, Permissible levels of toxic substances in the working environment, ILO, Geneva 1968.
[14] Malek, B., personal communication.
[15] Schneiderman, M. A., N. Mantel, C. C. Brown, From mouse to man – or how to get from the laboratory to Park avenue and 59th street., Ann. N. Y. Acad. Sci., **246** (1975) 237–248.
[16] Schöneich, J., Safety evaluation based on microbial assay procedures, International symposium on new developments in mutagenicity testing of environmental chemicals, Žinkovy Castle, 1975.
[17] Seltzer, J. R., Reaction grow to trichloroethylene alert, Chem. Eng. News, May 19 (1975) 41–43.
[18] Smyth, H. F., The toxicological basis of threshold limit values: 1. Experience with threshold limit values based on animal data, Amer. ind. Hyg. Assoc. J., **20** (1959) 341–345.
[19] IIeme Symposium international sur les limites tolérables pour les substances toxiques dans l'industrie, Institut national de sécurité, Paris 1963.
[20] United nations scientifiic committee on the effects of atomic radiation, ionizing radiation: levels and effects, 1972.
[21] von Hohenheim, T. B., (Paracelsus), 3. der 7. Kärtner Defensiones, 1537.
[22] WHO Scientific group, Assessment of the carcinogenicity and mutagenicity of chemicals, Technical Report Series 546, WHO, Geneva 1974.

The Significance of Long-Term Exposures to Chemical Mutagens in Mice

R. J. Šrám and V. S. Zhurkov

1) Institute of Hygiene and Epidemiology, 10042 Prague (Czechoslovakia)
2) Institute of Medical Genetics, Ac. Med. Sci. U.S.S.R., 115478 Moscow (U.S.S.R.)

Summary

A search through the literature has shown that there is still lack of experimental data analysing the time-effect relationship in long-term exposure to chemical mutagens. We have studied the time-effect relationship in mice chronically exposed to cyclophosphamide (CY), using bone marrow cytogenetic analysis in female mice exposed to 0.01 % CY for 1—70 days and dominant lethal test in male and female mice exposed for 1—12 weeks to 0.01 % and to 0.01 or 0.002 % CY respectively. Both bone marrow cytogenetic analysis and dominant lethal test have lead to a conclusion that chronic exposure induces detectable changes unrelated to the length of exposure. This finding remains to be verified for various mutagens, different routes of administration and for various mutagen doses, one of which corresponds to expected human exposure.

Long-term exposures to chemical mutagens and their risks for population health have become a subject of practical concern because there is a rapidly increasing number of people who are continually exposed to various chemicals some of which are potential mutagens.

In various schemes of mutagenicity testing the genetic risk estimation is usually based on an acute exposure to a single dose of mutagen, or in a better case on a subacute exposure to several repeated doses of tested chemical. The results are then extrapolated to expected or calculated long-term exposure values using various notions of their possible relationship.

What we need now is to find a generally acceptable relationship between the acute and chronic exposure allowing reliable estimate of future genetic risks on the basis of an acute exposure which is economically preferable to any other aproach that might be used in potential genetic risk estimation.

This relationship between acute and chronic exposures is of practial significience e.g. when analysing a potential genetic risk of alkylating agents in chemical industry, using the technique of cytogenetic analysis of peripheral lymphocytes of occupationally exposed workers. One may expect that the increase of aberrant cells should be related to the length of exposure, however, vinylchloride experience has not confirmed this presumption (Kučerová 1976).

On the basis of our previous experiments with mice exposed to TEPA, THIOTEPA, EMS and cyclophosphamide (CY) intraperitoneally applied in several repeated doses, we have come to a conclusion, using dominant lethal test and bone marrow cytogenetic analysis, that the present stage of our knowledge requires to accept an assumption, that the genetic risk of a dose totally received during a prolonged period is about the same as after a single acute exposure to the same total dose (Šrám 1976). Taking this approach and using the scheme of testing and evaluation recommended by Bochkov et al. (1976) we have estimated the potential genetic risk for workers occupationally exposed to epichlorhydrin (Šrám et al. 1976). As far as we know no such an attempt has been made to extrapolate practically the results from experimental models on possible human genetic changes due to an occupational exposure to industrial mutagens.

Do these new experimental results support this particular approach? Long-term exposures are generally believed to be the most suitable for assessing potential genetic risks, if applied correspondingly to the usual route of human exposure. Making search through the literature, we have found several long-term trials in mice or rats, however, these studies have been conduced with chemicals that are far from being considered strong mutagens e.g. 2,4-D, DDT, lindan, prometrin (Fomenko et al. 1973), or with chemicals, whose mutagenic activity has not been proved in mammals (as caffeine — Adler and Röhrborn 1969, Röhrborn 1972, DDVP — Dean and Thorpe 1972a, b, fenitrothion — Beneš et al. 1975). In addition to it, some results have been obtained in rats, which according to our present knowledge are less susceptible to chemical mutagens than e.g. mice. In some other cases, as e.g. in experiments with EI or chloropren (Fomenko et al. 1973) the effect of chronic exposure was found to be analysed only in few life-time intervals, which does not provide reliable basis for the assessment of its longitudinal tendency.

Neither with EMIC help were we able to find experimental data analysing the time-effect relationship of chemical mutagens.

So we are going to discuss here only the first experiences of our own in this field.

Our experiments were conducted on ICR random-bred mice, using bone marrow cytogenetic analysis and dominant lethal test in males and females. As a model mutagen we used Cyclophosphamide Germed (CY), cumulative effect of which after intraperitoneal application has been already observed (Šrám 1976). But in these experiments, drinking water-dissolved CY was applied perorally, i.e. correspondingly by to usual human exposure. CY was chosen because it is easy to reach and remains stable during normal storing conditions. Drinking water with CY was applied ad libitum, the samples of mutagen were prepared twice a week; the concentrations used were chosen according to screening test results.

Bone marrow cytogenetic analysis

Female mice aged 6—8 weeks were exposed to 0.01% CY for 1—70 days, each analysed group of mice consisted of 5 mice. Slides were prepared at the end of the defined exposure. 250 metaphases were scored per group, evaluating gaps, breaks and exchanges.

The frequency of aberrant cells detected in controls was 2.0 and 3.2%, with the average 2.6%. Percentage of aberrant cells in treated groups ranged depending on the length

Table 1

Chromosome aberrations in bone marrow cells of mice chronically exposed to 0.01 % cyclophosphamide in drinking water

Exposure days	No of mice	Total number of cells	Aberrant cells (%)	Types of aberrations (%)				B/C	G/C	Abnormal cells (%)
				single fragments	chromatid exchanges	double fragments	chromosome exchanges			
1	5	246	13.8	13.0	0.8	2.4	0.0	0.163	0.029	16.7
2	4	200	12.5	12.5	0.0	3.0	0.0	0.155	0.060	17.5
3	3	150	12.7	9.3	0.0	4.7	0.0	0.140	0.033	14.7
5	5	250	16.4	14.8	0.0	3.2	0.0	0.180	0.068	20.4
6	5	241	11.6	11.2	0.8	2.5	0.0	0.145	0.054	15.8
7	5	250	14.8	16.8	1.6	0.8	0.0	0.204	0.032	16.8
9	5	250	17.2	18.4	1.2	1.6	0.0	0.220	0.036	19.6
11	5	247	10.5	10.5	0.0	0.8	0.0	0.113	0.036	13.8
14	5	250	17.2	26.8	2.0	1.6	0.0	0.308	0.048	20.0
21	5	250	12.4	12.8	0.4	2.0	0.0	0.152	0.024	14.8
28	5	250	19.6	20.8	0.4	3.2	0.0	0.244	0.060	24.0
35	5	237	15.6	16.9	0.0	2.1	0.0	0.190	0.068	20.7
42	5	250	13.6	15.2	0.8	0.8	0.0	0.176	0.060	18.0
56	5	225	15.6	16.0	0.0	1.3	0.0	0.173	0.018	17.3
63	5	250	12.0	10.0	0.0	3.6	0.0	0.136	0.056	16.0
70	5	250	11.2	11.6	0.0	0.8	0.0	0.124	0.044	15.6
Control I	5	250	2.0	1.6	0.0	0.4	0.0	0.20	0.016	3.2
Control II	5	250	3.2	3.2	0.0	0.0	0.0	0.32	0.012	4.4
Control I + II	10	500	2.6	2.4	0.0	0.2	0.0	0.26	0.014	3.8

B/C, G/C — breaks, gaps per cell.

of exposure, from 10.5% (11 days) to 19.6% (28 days), the average frequency of aberrant cells being 14.2%, (Tab. 1, Fig. 1). Analysing the tendency of changes by χ^2 test, no difference was found between the individual groups ($\chi = 17.07$, $f = 15$, $p = 0.30$), i.e. the frequency of aberrant cells continually exposed to CY was not influenced by the length of exposure. Similar conclusion was also reached when analysing the yield of breaks per cell. The most significant category detected were single and double fragments. In some points there were also observed chromatid exchanges, mostly sister reunion (intrachromosomal assymetric exchanges). In order to exclude a chance that these results were only coincidental this experiment was repeated, but the tendency to balance the yields of aberrant cells was repeatedly by confirmed.

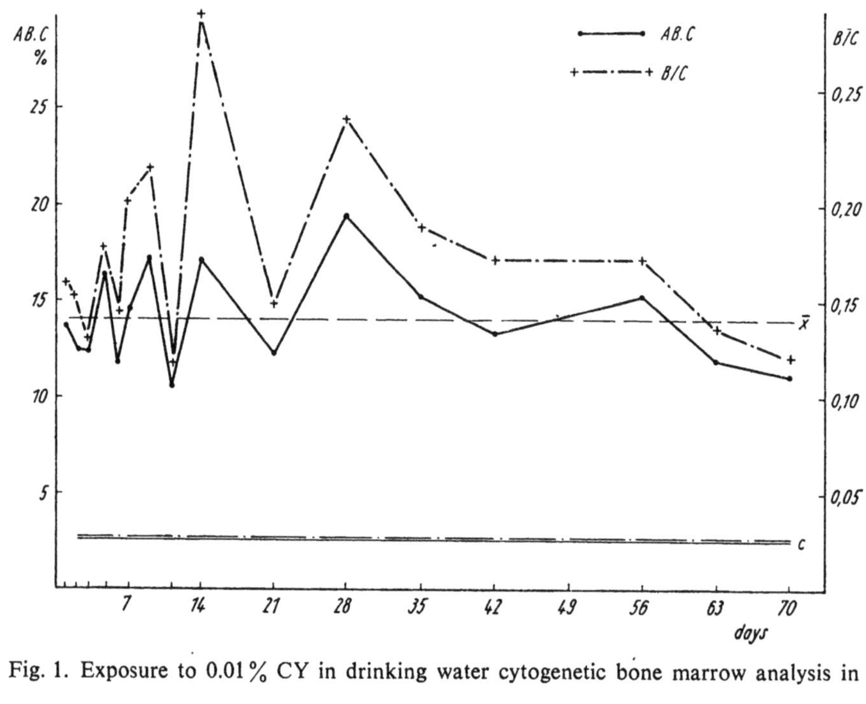

Fig. 1. Exposure to 0.01% CY in drinking water cytogenetic bone marrow analysis in mice.

These results show that the frequency of aberrant cells occurring in bone marrow of mice chronically exposed to 0.01% CY in drinking water does not exhibit changes dependable on the length of exposure.

Similar results were reported by PAYNE et al. (1974), who also failed to detect any changes in the frequency of aberrant cells in bone marrow of rats given for 5 generations food containing various doses of CY. It may be assumed that the scored frequency of aberrant cells is the result of a balance between the mutagen-induced production of aberrant cells and their natural elimination during mitotic division of bone marrow cells. Newly induced aberrant cells may also stimulate higher cell renewal, which may also help keep the detected changes on a potentially stable level.

Our conclusions are also backed by the fact that the main category of genetic injuries was represented by chromatid fragments, which were fastly eliminated.

These results differ from those obtained during a fractionated application. Probably one of the main reasons may be linked with the route of application: fractionated doses of CY were applied i.p. and CY-induced changes accumulated in affected cells. But continuous exposure to CY in drinking water resulted in nearly a continuous contact of mutagen with target cells, where we analysed a possible effect.

Dominant lethal test in males

Male mice were exposed to 0.01% CY for 1—12 weeks, exposure started at the age of 6—8 weeks. After exposure to CY, groups of 20 males were individually mated with 40 females in the course of 5 subsequent weeks. Dominant lethals were scored and calculated according to our usual scheme (Tab. 2), (Šrám et al. 1974).

Table 2

Dominant lethals induced by 0.01% cyclophosphamide in drinking water

Week after treatment	Pregnant females %	Corpora lutea/♀	Implantations/♀	Live embryos/♀	Dead embryos/♀	Total loss	Preimplantation loss	Postimplantation loss	DLM (%)
Exposure: 1 week									
1	95.0	13.1	10.7	8.3	2.4	0.3590	0.1723	0.2186	29.6
2	77.5	14.0	11.7	9.6	2.1	0.3144	0.1684	0.1811	17.1
3	85.0	13.6	11.3	10.2	1.1	0.2492	0.1648	0.1018	8.4
4	95.0	13.0	11.9	11.2	0.7	0.1485	0.0862	0.0677	− 3.2
5	77.5	13.8	10.9	10.4	0.5	0.2305	0.1914	0.0518	6.1
Exposure: 2 weeks									
1	85.0	13.4	10.6	6.1	4.5	0.5383	0.1994	0.4347	49.3
2	87.5	13.3	11.1	9.1	2.1	0.3071	0.1501	0.1837	16.3
3	92.5	13.9	11.5	10.9	0.6	0.2084	0.1668	0.0536	3.4
4	87.5	13.8	12.1	11.4	0.7	0.1660	0.1161	0.0582	− 1.0
5	82.5	13.6	11.4	11.0	0.5	0.1903	0.1558	0.0448	1.2
Exposure: 3 weeks									
1	90.0	13.7	11.6	8.9	2.7	0.3414	0.1462	0.2458	22.7
2	92.5	13.7	11.4	10.1	1.3	0.2610	0.1647	0.1238	10.7
3	95.0	14.1	12.2	11.4	0.7	0.1811	0.1247	0.0652	0.1
4	82.5	13.4	12.0	11.3	0.7	0.1533	0.1018	0.0562	− 2.6
5	72.5	14.2	11.8	10.9	0.8	0.2270	0.1673	0.0730	5.7
Exposure: 4 weeks									
1	57.5	14.9	10.1	7.0	3.2	0.5319	0.3116	0.3341	48.6
2	80.0	13.1	10.5	9.5	1.1	0.2741	0.1918	0.0975	12.3
3	92.5	13.0	10.9	10.3	0.6	0.2256	0.1772	0.0592	5.5
4	75.0	13.0	11.1	10.6	0.5	0.1768	0.1410	0.0402	0.2
5	77.5	13.7	11.4	10.9	0.5	0.1944	0.1562	0.0529	1.7

Table 2 (Continued)

Dominant lethals induced by 0.01% cyclophosphamide in drinking water

Week after treatment	Pregnant females %	Corpora lutea/♀	Implantations/♀	Live embryos/♀	Dead embryos/♀	Total loss	Pre-implantation loss	Post-implantation loss	DLM (%)
Exposure: 6 weeks									
1	67.5	13.9	11.2	9.3	1.9	0.3365	0.2027	0.1749	27.1
2	67.5	13.1	11.0	9.4	1.6	0.2976	0.1769	0.1440	15.1
3	72.5	13.7	12.1	11.6	0.5	0.1465	0.1166	0.0393	− 4.1
4	77.5	12.8	10.0	9.5	0.5	0.2686	0.2267	0.0754	11.4
5	72.5	14.1	11.7	11.3	0.5	0.1992	0.1671	0.0422	2.2
Exposure: 8 weeks									
1	70.0	13.3	9.9	7.8	2.1	0.4091	0.2434	0.2223	35.1
2	82.5	13.6	11.2	10.6	0.6	0.2137	0.1696	0.0544	5.0
3	92.5	13.5	11.8	11.3	0.5	0.1534	0.1199	0.0382	− 3.2
4	70.0	13.6	11.3	11.0	0.3	0.1983	0.1753	0.0434	2.9
5	57.5	12.7	10.6	9.8	0.8	0.2346	0.1759	0.0602	6.6
Exposure: 10 weeks									
1	77.5	13.9	11.1	9.4	1.7	0.3175	0.1928	0.1623	25.0
2	65.0	13.0	11.3	10.5	0.8	0.1921	0.1308	0.0721	2.4
3	80.0	13.9	11.7	11.2	0.5	0.1899	0.1579	0.0445	1.2
4	85.0	13.4	11.2	10.6	0.6	0.2097	0.1655	0.0641	4.2
5	87.5	14.5	13.0	12.4	0.5	0.1353	0.0977	0.0424	− 5.5
Exposure: 12 weeks									
1	47.5	13.2	10.7	7.2	3.6	0.4583	0.1814	0.3602	40.5
2	55.0	12.0	9.9	8.7	1.2	0.2723	0.1678	0.1268	12.1
3	62.5	12.0	8.3	7.8	0.5	0.3634	0.3203	0.0640	22.3
4	55.0	13.0	10.8	10.2	0.7	0.2345	0.1819	0.0608	7.2
5	35.0	14.1	10.1	9.5	0.6	0.3285	0.2839	0.0978	18.1
Control:									
1	77.5	13.7	12.9	12.4	0.5	0.0896	0.0549	0.0363	
2	75.0	12.7	11.2	10.5	0.7	0.1724	0.1154	0.0739	
3	85.0	13.1	11.2	10.7	0.6	0.1802	0.1372	0.0471	
4	95.0	14.1	12.0	11.6	0.4	0.1747	0.1441	0.0355	
5	87.5	14.2	12.0	11.5	0.5	0.1806	0.1455	0.0398	

No changes in male fertility were evidenced during long-term exposure, only in the group exposed for 12 weeks the frequency of pregnant females significantly decreased in the 1st, and 3rd to 5th week (Fig. 2).

In all exposed groups of mice the most significant effect was observed during the first week after stopping exposure (Fig. 3). Total and postimplanation losses were also increased in week 2 in males exposed for one and two weeks. This is why the time-effect relationship was analysed in this particular period and genetic injuries were calculated either as total and postimplantation losses (Fig. 4), or lethal events and dead embryos per female (Fig. 5). The highest frequency rate was found after the exposure for two or four weeks, but these changes were found to be time-independent.

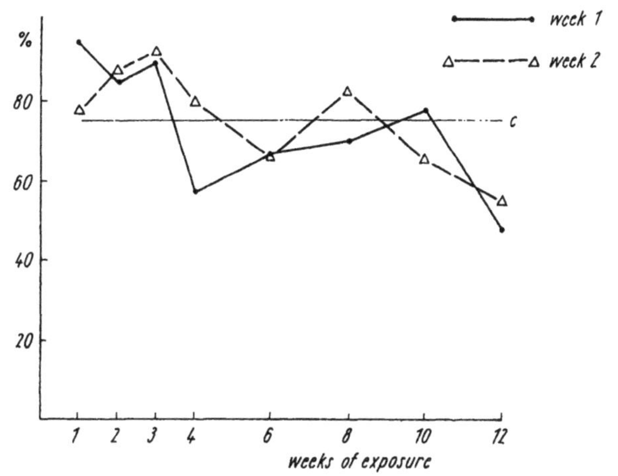

Fig. 2. Exposure to 0.01 % CY in drinking water; dominant lethals in male mice; percentage of pregnant females.

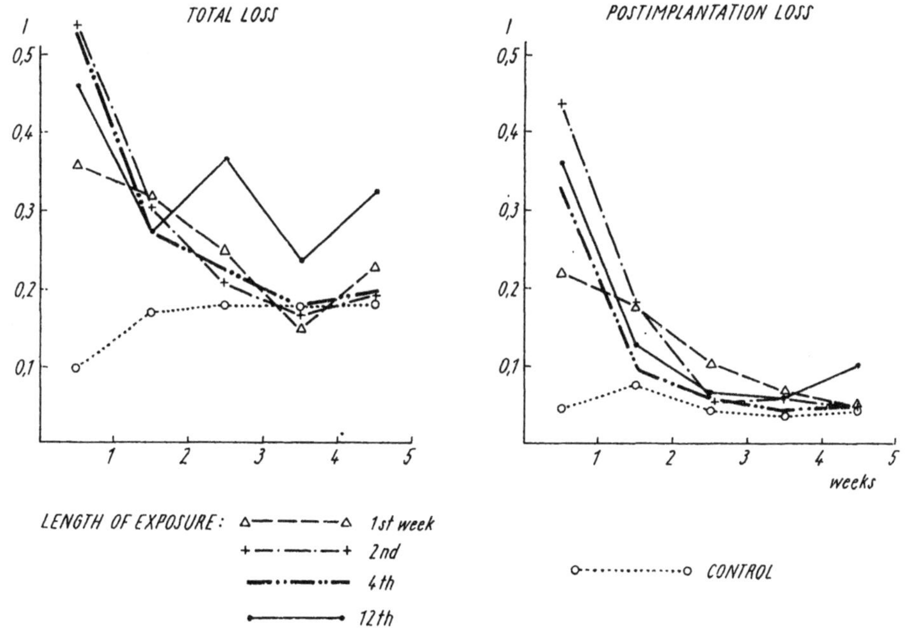

Fig. 3. Exposure to 0.01 % CY in drinking water; dominant lethals in male mice.

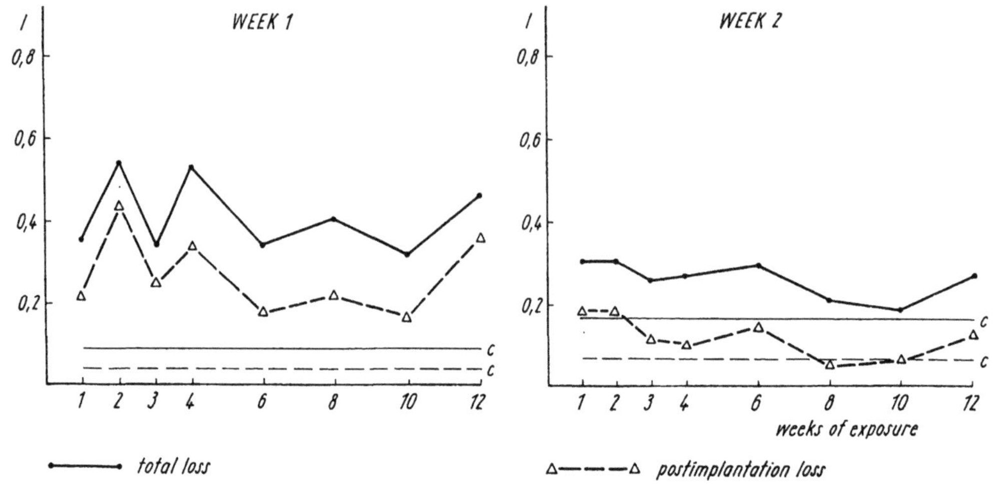

Fig. 4. Exposure to 0.01% CY in drinking water; dominant lethals in male mice.

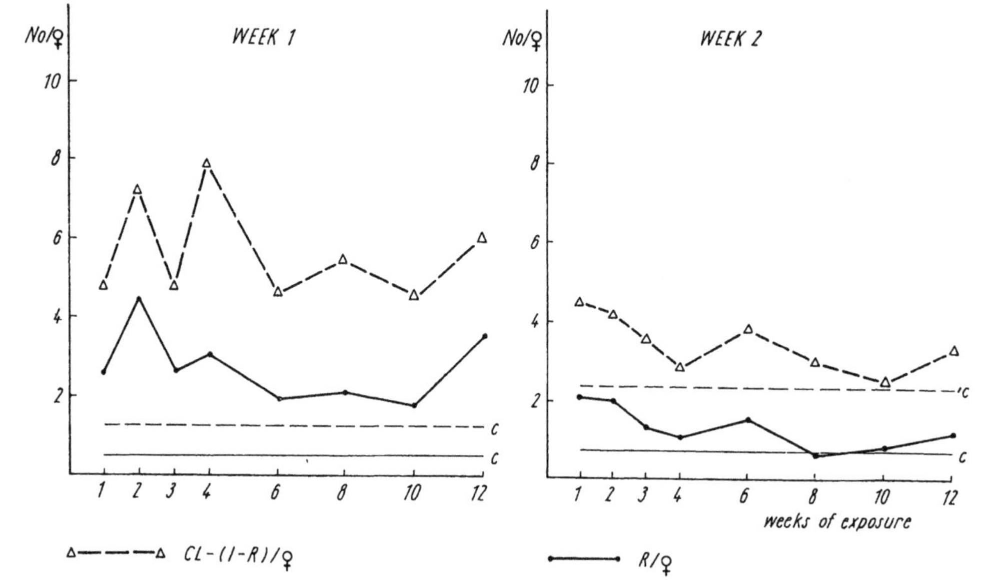

Fig. 5. Exposure to 0.01% CY in drinking water; dominant lethals in male mice.

The frequency of dominant lethals scored after long-term exposure was found to differ from a general pattern, if CY was applied in a single dose. In our present experiments the most significant effect, and in some groups also the only one, was detected during the first week after stopping the exposure to mutagen. However, in some treated groups, CY acted on all spermatogenic stages. The most significant results observed only during the first week may be attributed to a strong renewal in the stage of spermatogonia and spermatocytes, when cells bearing chromosome aberrations are unable to go through the meiotic division.

The results of both dominant lethal test and bone marrow cytogenetic analysis concordantly indicate that the genetic injury in mice do not change with the length of exposure to mutagen. It is not surprising if we take into consideration that chromosome aberrations in gametes are the main source of dominant lethals. Analogously to bone marrow, the scored effect of long-term exposure to CY in drinking water resulted from the balance between the induction of aberrant cells and their elimination, in this case mainly during meiotic division.

From this aspect the mutagenic effect on postmeiotic cells may be considered as the most significant. For practical purposes of mutagenicity testing it may also lead to a conclusion, that time-limited exposure is sufficient for a general appraisal of the genetic risks of long-term exposure to mutagen. But such a conclusion still remains to be verified by testing several mutagens at various routes of exposure, especially at peroral and inhalation applications.

Dominant lethal test in females

Female mice were exposed to 0.002 and 0.01% CY for 1—12 weeks, exposure started at the age of 6—8 weeks. After stopping exposure to CY, groups of 40 females were mated with 20 males (one male with two females) in the course of the 1st, 2nd and 5th subsequent week (Tab. 3—7).

Table 3

Dominant lethals induced by cyclophosphamide in drinking water (females exposed for 1 week)

Week after treatment	Pregnant females %	Corpora lutea/♀	Implantations/♀	Live embryos/♀	Dead embryos/♀	Total loss	Pre-implantation loss	Post-implantation loss	DLM (%)
Dose: 0.002% CY									
1	80.0	13.2	12.6	11.7	0.9	0.1170	0.0505	0.0688	− 4.3
2	70.0	13.0	12.4	11.7	0.7	0.1059	0.0512	0.0578	−15.7
5	62.0	14.5	13.9	13.1	0.7	0.0911	0.0412	0.0516	− 2.0
Dose: 0.01% CY									
1	70.0	12.2	11.2	10.5	0.6	0.1437	0.0906	0.0596	− 1.1
2	44.0	11.8	11.1	10.5	0.5	0.1124	0.0597	0.0612	−14.9
5	74.0	14.3	13.7	12.9	0.7	0.1043	0.0478	0.0571	− 0.5
Control:									
1	70.0	12.6	11.8	10.8	0.9	0.1535	0.0675	0.0952	
2	46.0	12.0	10.8	9.5	1.3	0.2276	0.0991	0.1318	
5	62.0	13.3	12.5	12.0	0.4	0.1090	0.0738	0.0380	

Table 4

Dominant lethals induced by cyclophosphamide in drinking water (females exposed for 2 weeks)

Week after treatment	Pregnant females %	Corpora lutea/♀	Implantations/♀	Live embryos/♀	Dead embryos/♀	Total loss	Pre-implantation loss	Post-implantation loss	DLM (%)
Dose: 0.002% CY									
1	84.0	14.7	13.7	13.2	0.5	0.0979	0.0628	0.0375	− 8.3
2	92.0	14.6	13.2	12.8	0.3	0.1177	0.0903	0.0315	− 9.4
5	54.0	13.0	12.1	10.7	1.3	0.1666	0.0692	0.1028	5.7
Dose: 0.01% CY									
1	84.0	13.6	12.1	11.5	0.5	0.1504	0.1074	0.0551	− 1.9
2	82.0	14.2	12.5	11.9	0.5	0.1732	0.1281	0.0513	− 2.5
5	84.0	13.1	11.7	11.3	0.4	0.1382	0.1029	0.0399	2.5
Control:									
1	80.0	15.6	13.5	12.9	0.6	0.1670	0.1271	0.0484	
2	70.0	14.7	12.0	11.6	0.4	0.1941	0.1599	0.0467	
5	60.0	13.0	12.3	11.6	0.7	0.1157	0.0610	0.0581	

Table 5

Dominant lethals induced by cyclophosphamide in drinking water (females exposed for 4 weeks)

Week after treatment	Pregnant females %	Corpora lutea/♀	Implantations/♀	Live embryos/♀	Dead embryos/♀	Total loss	Pre-implantation loss	Post-implantation loss	DLM (%)
Dose: 0.002% CY									
1	82.0	12.6	11.4	10.8	0.6	0.1447	0.0959	0.0561	3.9
2	80.0	13.6	12.0	11.6	0.4	0.1449	0.1186	0.0436	− 1.5
5	72.0	14.1	11.9	11.4	0.4	0.1908	0.1568	0.0430	1.2
Dose: 0.01% CY									
1	78.0	12.4	10.9	10.0	0.8	0.1867	0.1180	0.0940	8.6
2	84.0	13.3	11.4	10.7	0.6	0.1701	0.1175	0.0567	0.9
5	88.0	13.7	12.0	10.9	1.0	0.2053	0.1328	0.0902	2.9
Control:									
1	76.0	13.1	12.1	11.7	0.4	0.1102	0.0789	0.0331	
2	78.0	13.7	12.0	11.5	0.5	0.1618	0.1228	0.0451	
5	94.0	14.3	12.1	11.7	0.3	0.1810	0.1546	0.0321	

Table 6

Dominant lethals induced by cyclophosphamide in drinking water (females exposed for 8 weeks)

Week after treatment	Pregnant females %	Corpora lutea/♀	Implantations/♀	Live embryos/♀	Dead embryos/♀	Total loss	Pre-implantation loss	Post-implantation loss	DLM (%)
Dose: 0.002% CY									
1	82.0	14.2	11.9	11.4	0.5	0.2059	0.1680	0.0437	8.4
2	94.0	14.0	12.1	11.6	0.4	0.1727	0.1374	0.0399	− 6.9
5	72.0	13.5	11.7	11.4	0.3	0.1526	0.1277	0.0274	0.9
Dose: 0.01% CY									
1	86.0	13.1	11.5	10.5	1.0	0.2044	0.1244	0.0886	8.2
2	90.0	12.6	10.9	10.2	0.6	0.1861	0.1324	0.0621	− 5.2
5	84.0	13.0	11.2	10.4	0.7	0.1995	0.1414	0.0682	6.4
Control:									
1	78.0	14.1	13.0	12.2	0.7	0.1329	0.0789	0.0625	
2	86.0	14.1	11.6	11.0	0.6	0.2268	0.1785	0.0607	
5	90.0	14.1	12.9	12.2	0.7	0.1445	0.0912	0.0582	

Table 7

Dominant lethals induced by cyclophosphamide in drinking water (females exposed for 12 weeks)

Week after treatment	Pregnant females %	Corpora lutea/♀	Implantations/♀	Live embryos/♀	Dead embryos/♀	Total loss	Pre-implantation loss	Post-implantation loss	DLM (%)
Dose: 0.002% CY									
1	78.0	12.8	12.0	11.6	0.3	0.0985	0.0686	0.0305	− 1.4
2	70.0	12.6	11.6	11.0	0.5	0.1316	0.0862	0.0645	− 5.2
5	56.0	13.8	13.0	12.2	0.7	0.1135	0.0620	0.0539	− 7.6
Dose: 0.01% CY									
1	80.0	11.0	10.0	8.8	1.1	0.2107	0.0973	0.1292	11.2
2	92.0	13.3	12.1	11.3	0.8	0.1597	0.0973	0.0732	− 1.8
5	70.0	12.4	10.8	10.4	0.4	0.1711	0.1345	0.0454	− 0.6
Control:									
1	82.0	13.8	13.1	12.3	0.7	0.1113	0.0531	0.0620	
2	80.0	13.8	12.2	11.5	0.7	0.1744	0.1189	0.0650	
5	48.0	13.6	12.2	11.3	0.8	0.1758	0.1165	0.0657	

The total loss was found to be increased in the 1st week after the exposure to 0.01% CY applied for 4,8 or 12 weeks. All other groups did not differ from the control. No changes in female fertility were detected in any experimental group of mice (Fig. 6).

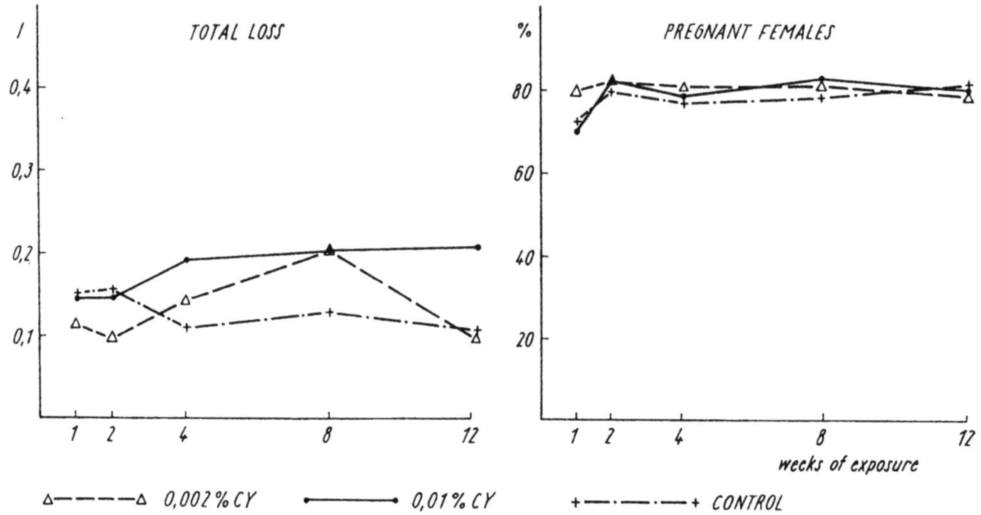

Fig. 6. Exposure to CY in drinking water, dominant lethals in female mice (week 1)

Analysis of dominant lethals in females has confirmed hypothetical assumption, that chromosome aberrations may increase the frequency of scored lethals, if they are induced predominantly in postmeiotic stage. It has been already cytogenetically proved that chromosome aberrations may be induced also in oocytes, however, it is probably very difficult for this oocyte to be fertilized and become a normally developing zygote. Dominant lethal test is therefore less sensitive in female mice than in males.

Being already in dictyotene, oocytes may be affected by mutagen for a longer period of time. As these cells, in contradistinction to male spermatogonia and spermatocytes, are not going through the phase of mitotic division, genetic changes may, therefore, accumulate in individual oocytes. Dominant lethals in females may result also from nondisjunction. But these hypothetical possibilities are to be proved cytogenetically in one of the next experiments.

Conclusion

Our experimental data presented here are to be understood as the first steps attempting to study long-term exposures to mutagens.

However, we believe that our experiments with CY in drinking water may already lead to some conclusion.

Using bone marrow cytogenetic analysis and dominant lethal test in mice we have found out that long-term exposure induces a stable frequency of scored changes unrelated to the length of exposure.

Continuous application of mutagen, which will most probably change also the toxicological impact of mutagen on the studied organism, may reveal a more complex response of the model organism than tests with acute or fractionated exposures. Since the most significant injury detected during spermatogenesis was found to be induced in postmeiotic stage it might be possibly assumed that the exposure corresponding to the length of this period may yield basic information about the genetic risk of long-term exposures to mutagens. Compared to an acute exposure, this approach requires much lower doses to be applied, which however, affect the target cells continually. It means that we may analyse a cumulative effect of these lower doses thus compensating a higher single dose. But it is to be at the same time understood that the scored mutagenic activity is the result of a balance between the newly induced chromosome aberrations and their natural elimination during the cell cycle.

According to our present knowledge, chemical mutagens inducing chromosome aberrations are capable of producing also gene mutations which are difficult to eliminate and which may coincidentally occur in the same cells bearing chromosome aberrations. Such a situation characterized by a stable level of chromosome aberrations and by an accumulation of gene mutations in the analysed tissue may be, at least for a certain period of time, irrelevant for human health, however, in the long run it may stimulate the development of malignant clones. From this particular point of view, microquantites of chemical mutagens are significant not only because of their genetic risks for next progeny, but also because of the risk of increased somatic mutations in exposed persons.

In future genetic studies concerned with long-term exposure to chemical mutagens it will be recommendable to take into consideration also the following aspects:

1) In order to make a better assessment of the genetic risks it will be perhaps necessary to use cytogenetic methods facilitating to score the frequency of mutations also in cells requiring a relatively long period for their renewal, as e.g. peripheral lymphocytes, liver and brain cells, oocytes, or methods scoring gene mutations. These methods should allow to score or estimate the frequency of mutations induced during the whole exposure period.

The methods which are now routinely used, as bone marrow cytogenetic analysis or dominant lethal test, are less suitable for this particular purpose. The most promising seems to be the cytogenetic analysis of chromosome aberrations in mammalian liver cells, analysing not only normally dividing cells but also cells that participate in the biotransformation of all the chemicals introduced into the organism.

2) The effect of long-term exposure to mutagens should be analysed at different routes of administration, particularly after peroral application on drinking water and food or following inhalation exposure, and the mutagenic activity should be tested for various doses, one of which corresponds to human exposure.

3) Studies concerned with the effect of low doses of chemical mutagens may facilitate to reach or calculate the level, which — at the present sensitivity of available methods — corresponds to the level of spontaneous mutations. This level might be then accepted as the maxim allowable concentration (MAC) for the particular chemical.

4) The results of dominant lethal tests in male and female mice seem to indicate that the impacts of chemical mutagens might be more serious in males than in females. This should be kept in mind if practical measures are to be taken in those occupationally exposed to industrial chemicals.

Acknowledgement

The authors thank Dr. J. S. Wassom, Environmental Mutagen Information Center, Oak Ridge National Laboratory, U.S.A., for providing the search of literature.

References

[1] ADLER, I. D. and G. RÖHRBORN, Cytogenetic investigation of meiotic chromosomes of male mice after chronic caffeine treatment. Humangenetik, **8** (1969) 81–85.

[2] BENEŠ, V., R. J. ŠRÁM and R. TUSCANY, Fenitrothion. I. Study of mutagenic activity in rats. J. Hyg. Epidemiol. Microbiol. Immunol. (Praha), **19** (1975) 163–172.

[3] BOCHKOV, N. P., R. J. ŠRÁM, N. P. KULESHOV and V. S. ZHURKOV, System for the evaluation of the risk from chemical mutagens for man: basic principles and practical recommendations. Mutation Res., **38** (1976) 191–201.

[4] DEAN, B. J. and E. THORPE, Cytogenetic studies with dichlorvos in mice and Chinese hamsters. Arch. Toxikol., **30** (1972a) 39–49.

[5] DEAN, B. J. and E. THORPE, Studies with dichlorvos vapour in dominant lethal mutation tests in mice. Arch. Toxikol., **30** (1972b) 51–59.

[6] FOMENKO, V. N., E. E. STREKALOVA, L. D. KATOSOVA, E. M. CHIRKOVA, L. S. SAĹNIKOVA, I. V. SILANTEVA, L. P. EFIMENKO and A. E. KULAKOV, Some experimental results on adaptation and its limitations due to effect of chemicals with mutagenic and embryotropic activity (In Russian). In: Pharmacology. Chemotherapeutical Drugs. Toxicology. – Problems in Toxicology. Vol. 5. Moscow 1973, p. 128–145.

[7] KUČEROVÁ, M., Cytogenetic analysis of human chromosomes and its value for the estimation of genetic risk. Mutation Res., (1976) in press.

[8] PAYNE, H. G., L. W. NELSON, and J. H. WEIKEL, Effect of cyclophosphamide on somatic cell chromosomes in rats. Toxicol. Appl. Pharmacol., **30** (1974) 360–368.

[9] RÖHRBORN, G., Mutagenitätsuntersuchungen an Mäusen nach chronischer Behandlung mit Coffein. Z. Ernährungswissenschaft (Suppl.), **14** (1972) 54–67.

[10] ŠRÁM, R. J., Relationship between acute and chronic exposures in mutagenicity studies in mice. Mutation Res., (1976) in press.

[11] ŠRÁM, R. J., M. ČERNÀ and M. KUČEROVÁ, The genetic risk of epichlorhydrin as related to the occupational exposure. Biol. Zbl., **95** (1976) 451–462.

[12] ŠRÁM, R. J., Z. ZUDOVÁ and P. GOETZ, The mutagenic effect of lysergic acid diethylamide. II. Dominant lethal test in mice. Mutation Res., **26** (1974) 517–522.

Symposium III

Population Exposure and Genetic Monitoring

Chairman's Introduction

Per OFTEDAL

The proof of all puddings comes — as the saying goes — with the eating. No amount of good raw material and fancy cooking is of any help if the result is not a palatable disk.

In environmental mutagenesis, the proof would appear to come from the demonstration of effects in exposed human populations. This is certainly true, and a positive finding of transmitted genetic effects is the ultimate climax of a case analysis. Therefore, and in spite of all the difficulties in methods, in criteria, in politics, and in interpretation, it is extremely important that suspect situations should be analyzed in depth, and that efforts should be made even when they are costly and the prospect of a positive finding appears uncertain or slight.

Our symposium today is concerned with these matters: Population exposure and genetic monitoring. We have speakers on base line criteria, on experimental studies with the purpose of identifying threatening substances in the industrial environment, and on the measurement of effects in exposed human populations.

There is no longer discussion on the reality of the dangers of chemical mutagens in the environment. There is now a real need for the development of criteria and techniques for a practical and realistic approach to monitoring of selected human populations.

An important object of exercises of this kind is — apart from the primary one of protection of the public in that particular situation — to provide a scale: to demonstrate in one given situation that there is or is not an effect, so that this scale value — or limit — can be used to evaluate effects and risks to be expected in other situations, where only part of the scene is lit up and can be observed clearly. This may be because the effects cannot be identified or measured under the circumstances, or because it is a matter of projection into the future.

For this is where environmental mutagenesis differs from cooking: generally — and hopefully — there will be and should be no pudding to taste. If the meal is on the table, we are already too late, and have failed in our intentions.

The allegory should not be elaborated too far. But it is obvious that in real life a very great number of uncertainties will have to be accepted in the process of calculating genetic risks. It is also clear that such uncertainties leave room for each of us to lay emphasis on the various elements in an individual way, according to our own special interests and tastes. A theoretical population geneticist — or a chemist — may evaluate a problem quite differently from a clinical geneticist — or an experimental toxicologist — to mention a few possibilities.

But these are difficulties that can be overcome. Most easily by accepting each and everyone's preferences in this field and summing them all, or — which is more difficult, and where symposia like this are important — by discussion of views and evaluation of the importance of the various elements in the more relaxed and less compulsory atmosphere of the scientific meeting.

So the main difficulties remain in the subject matter itself: the small population, the exposure criteria, the identification of damage. Allow me to give one example which demonstrates some of the complexities, limitations and possibilities in this type of situation:

A year ago — at Zinkovy — I reported on the investigations in progress — slow progress — in the surroundings of a Norwegian vinyl chloride polymerization plant. We did find from the data in the Norwegian Medical Birth Registry that the frequency on anencephaly was somewhat elevated in the community. But with about 500 births per year in the town, the absolute numbers were too small to decide if the increase was real. However, in view of the long lifetime of VC in the atmosphere when sunlight is absent, there was a possibility for a seasonal effect. The number of cases in the local situations was again — of course — too small for any such analysis, but it was obviously necessary to see if a seasonal variation in anencephaly in Norway in general was present. The Medical Birth Registry has data for the years 1967—1974, and for about 65000 births per year in the whole country, the numbers permit a monthly analysis. The raw numbers are shown in the first slide — from 12 to 30 per month — and the rates — since birth number varies with the season — in the second slide. It will be seen that the rate varies through the year, and that there is an abrupt change of about fourfold from May to August, from .2 to .8 per thousand live births (Fig. 1).

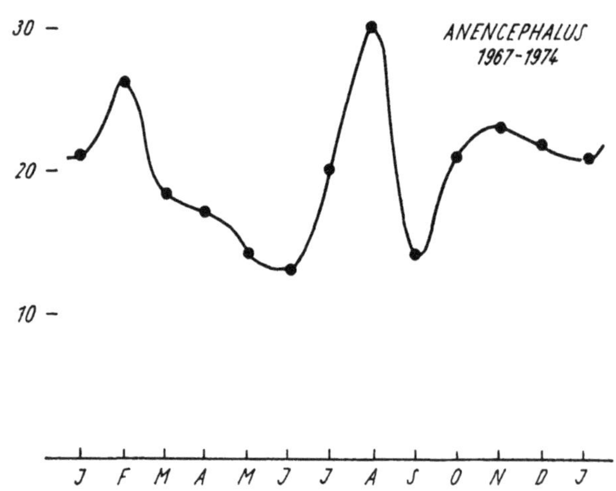

Fig. 1. Monthly number of cases of anencephalus amongst live births in Norway during 1967—1974. Data from the Norwegian Medical Birth Registry, University of Bergen.

No definite interpretation can be offered. The period of observation is too short to tell if this is a recurrent phenomenon or a changing situation. The data have not been looked at in terms of regions of the country, or urban/rural situation etc. It may be that Norwegian women do something — leading to anencephaly — in December (7—8 months before the peak) that they don't do in October (7—8 months before the through). Or it may be that anencephalous fetuses are aborted differentially according to season, and that may be Norwegian women do something — influencing abortion — in October, that they don't do in December. Thus, the question of teratogenicity as genetic effect is still open, though the abrupt variation would normally suggest a teratogenic etiology, and not a genetic one. The complexity of the matter is not reduced by the facts that may be 90% of anencephalous cases die before birth and are reported as monstrous abortions, and that the frequency of these abortions — as reported in earlier years, — show no seasonal variation (Fig. 2).

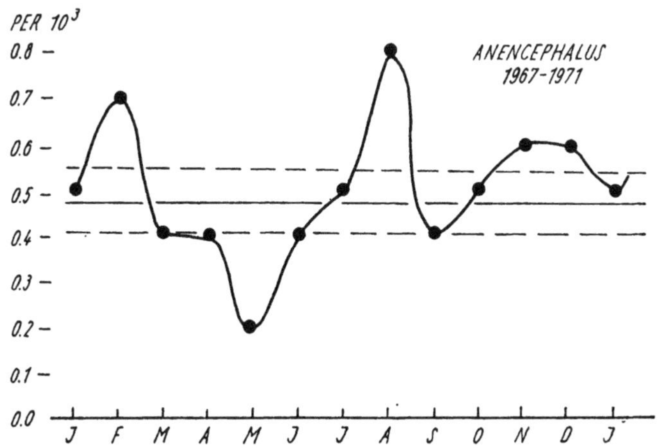

Fig. 2. Monthly rate of anencephalus per 1000 live births in Norway during 1967—1971. Mean ±2 s.d. Data from the Norwegian Medical Birth Registry, University of Bergen.

I think this example is a good illustration of some of the problems which are now going to be discussed by Dr. Barrai. The title of his talk is "Spontaneous mutation rates in man".

Spontaneous Mutation in Man*

I. Barrai

Department of Zoology, University of Ferrara
Ferrara, Italy

1. Introduction

1.1 Definitions

Mutation is a general property of all biological material. Specifically, it refers to the process resulting in transmissible changes of the hereditary material; at the limit, such changes may appear at the single locus and affect just one base pair inside a cistron, or involve complete chromosome sets and result in polyploidy. Several classifications have been developed to categorize mutations of different type. One classification is based on the size of the genomic material involved in the mutational event, so that the dimensional classification is consistent with the anatomical discontinuity of the genetic material.

1.2 Point mutations

So defined are those changes which involve one base pair inside a cistron; when the immediate translation product is recognized, as in haemoglobin, it is possible to infer with some accuracy the individual base pair involved in the change. To the class of point mutations one may attribue also those intracistronic changes, such as deletions, duplications or additions which segregate as simple Mendelian traits.

1.3 Chromosome mutations

To the class of chromosome mutations belong the various deletions, duplications, inversions and translocations of chromosome segments. The advent of the banding techniques has made possible the detection of a number of small inversions, translocations, and other fine structural abnormalities which are reported to be correlated with sterility and repeated abortion.

---

* This paper was prepared while the author was a visiting fellow of the WHO Collaborating Centre for Reference in Processing of Human Genetics Data at the Population Genetics Laboratory, University of Hawaii.

1.4 Aneuploidies

This class of mutations refers to the loss or addition of entire chromosomes in a genome. To the class belong monosomies, trisomies, and polysomies of the sex chromosomes. In man, the spontaneous frequency of some aneuploidies which are compatible with life has been studied with some detail.

1.5 Polyploidies

The addition of complete chromosome sets has been observed in man, but it was never compatible with life.

1.6 Origin of mutations

Mutations may be classified according to their origin, and in this case the main separation refers to spontaneous and induced mutations. However, spontaneous mutations are so defined because we ignore the causes which have induced them; in structure, spontaneous and induced mutations seem to be the same, indicating the same reaction of the hereditary material to known and unknown mutator accidents, be they in the form of chemical or physical mutagens. In most experimental organisms, the types of mutation observed after induction are the same as the types of mutations observed in wild type populations of the same organism. This may be true also for man; in the population groups which have been exposed to the action of known mutagens, no new types of changes in karyotypic or phenotypic effects have been reported.

2. The recognition of mutations

2.1 Detection through a phenotype

In man, a multitude of mutations has been described. The last edition of the catalog produced by McKusick lists more than 1850 entities corresponding to as many mutations which are transmitted autosomally or through the sex chromosomes, with dominant, or recessive effects. Most of the mutations described are either deleterious or lethal, and a few may be considered neutral. In fact, most human mutations are detected at the clinical level, because they produce an effect which is reported and recorded, and therefore may be at some stage the object of description and of counting. As a rule, mutations in man are described by their corresponding phenotype; they may be detected in the perinatal period, as many aneuploidies are, or later in life, depending on the fact that the effect of the mutation is expressed early during intrauterine life, or at later stages of the life cycle, particularly if the mutation is causing a disease characterized by an age of onset. At this point, it seems convenient to stress that most mutations are indeed deleterious, and contribute to human disease. We do not entertain any doubt that an increase in mutation rates would result in an increase in morbidity of genetic origin.

2.2 Detection at the clinical level

Among the traits listed by McKusick in this catalog, there are about 900 dominant, 800 recessive and 150 sex-linked. The vast majority of these traits is constituted by diseases, and among these most are lethal. It would then seem obvious that the statement closing the previous paragraph does represent a reasoned conviction and not an educated opinion. The opinion may be entertained on deductive grounds: since an allele is an ordered information system with an extremely high degree of canalization, any random noise introduced in it has a high chance of altering for the worse the outcome of the system. As a consequence, it would seem efficient to look for mutations in series of hospital patients, and to look for rates of mutations in those vital statistics systems which describe morbidity and mortality according to well defined codes based on the diagnoses determined by the physicians and reported to public health authorities.

3. The measurement of mutation rates

3.1 Direct and indirect method

There are two main methods of estimating mutation rates. The first, or direct method, is applicable to dominant traits, and requires the counting of sporadic cases with exclusion of those having a familial origin in a series of births. The estimator of the rate of mutation is simply the ratio between the number of sporadic cases and the total births. This method appears as a simple one, and has been applied for the estimation of several dominant rates. It is however subject to several sources of error, such as incomplete penetrance, phenocopies, non-paternity. The indirect method takes into account the relative fitness of the trait, and is applicable to both dominant and recessive traits, both autosomal and sex-linked. It is based on the assumption that the genes lost because of diminshed fitness are replaced by mutation; therefore, it is strictly applicable only in equilibrium populations. Most mutation rates of recessive and sex-linked have been estimated with the indirect method. Other methods have been devised and used efficiently, particularly by the school of MORTON (DEWEY et al., 1965). These methods are based on the theory of genetic loads proposed by MORTON et al. (1956) and further developed and applied by MORTON . These methods require that the level of inbreeding of individuals is also ascertained, and, although very powerful, they have not been generally applied. They have been most useful in the estimation of cumulative mutation rates at several loci.

3.2 Some considerations about mutation rates

Spontaneous rates of mutation may be due to environmental factors which are, as yet, unidentified. There is no doubt that cosmic and natural radiation are mutagenic factors; however, no correlation has been found between the intensity of cosmic radiation and mutation rates in man. SLATIS (1953) studied seasonal variation of sex ratios which may be correlated with the seasonal variation of the intensity of radiation, but too many variables affect sex ratio, and such studies are, as yet, inconclusive. Were cosmic radiation and the natural radiation important factors in determining mutation, one should be able to observe variation of spontaneous mutation with geographic factors. Studies in Minas Geraes and Kerala have indicated that this might not be the case in man, and one might accept the working hypothesis that mutation rates are constant in space. Again, the evidence comes from the works done by FREIRE-MAIA in Brasil and by SINISCALCO in India. The assumption of a constant mutation rate in space is the same as the assumption that there is not significant variation of the environmental factors which cause spontaneous mutation. Under the hypothesis that mutation rates are constant in space, one might also derive as a corollary that the rates are constant in time, if no change of environment is involved. This is practically the hypothesis from which originates our concern, because we are now observing rapid changes in our environment, namely in the number and quality of chemical products which may come in close contact with our germ plasm. The only acceptable variation for mutation rates would seem the variation observed at different loci of the same organism, and also the variation between organisms. Again, for point mutations one might expect that there is no significant variation of rates between loci, since the anatomy of the DNA is fairly constant, and the size of genes is not highly variable, and point mutation is a property of the DNA and not a property of the locus. However, we are able to detect only the phenotypic effect of mutation, and we would hardly be able to detect a constancy of rates, given the variability of phenotypic effects. For a closer examination of the assumptions, it is convenient to study the variation of estimates of mutation rates at the same locus in different human groups. In the following table, (Tab. 1), the ranges of estimates for

Table 1

Trait	Mutation per million gametes		
	Minimum (m)	Maximum (M)	Ratio M/m
Diaphyseal aclasis	6	9	1.5
Epiloia	8	16	2.0
Marfan's syndrome	4.2	5.8	1.4
Osteogenesis imperfecta	7	13	1.8
Retinoblastoma	7	46	6.6
Duchenne muscular dystrophy	43	95	2.2
Facioscapulohumeral muscular dystrophy	0.5	5	10.0
Haemophilia A	13	57	4.4
Hypogammaglobulinemia	3	5	1.7
Achondroplasia	10	13	1.3
Polycystic disease of the kidney	65	120	1.8
Average	15	35	3.2

several loci are given. The highest range is covered by facioscapulohumeral muscular dystrophy, which goes from a minimum of .5 mutants per million gametes (MORTON and CHUNG, 1959), to a maximum of 5 mutants per million, with a tenfold variation. The more stable estimates are those for achondroplasia, where the ratio between the maximum and minimum estimates is 1.3, giving substantial identity. In average, over the eleven loci for which we had data, there is a variation of more than three times between the maximum and minimum estimate of the same mutation rate.

3.3 Chromosome aberrations

Chromosome variations revealed by banding techniques are much more frequent than rates of mutation of individual genes. It would be a point in question if the frequency of a given detectable variation is more or less constant than mutation at a single locus; since chromosomal variations are studied as classes of variation, they may represent cumulative rates. Aneuploidies are even more frequent than chromosome mutations, and they show considerable stability of rates between populations (JACOBS, 1976). STEVENSON seems to have detected an interesting exception in a series of births from India, where trisomy 21 was exceedingly rare. In general, the ranges of frequencies of aneuploidies are much narrower than the ranges of point mutation rates, possibly because simulation of the same phenotype is rare, or because the higher frequency permits more precise estimates of the rates. These are typically 200 times larger than rates of point mutation. The question then arises, if the monitoring of aneuploidies, either at birth or in abortion material, is not more important than monitoring for point mutation. Polyploids represent a special case, in the sense that their frequency is rare, and none has been observed to be compatible with life.

4. Problem of choice of "spy mutants" or "sentinel phenotypes"

4.1 The indicator phenotype

A problem which arises immediately in the study of monitoring, refers to the choice of an indicator phenotype. As far as spy mutants are concerned, VOGEL in 1970 listed 5 markers which might be used for monitoring. Among these retinoblastoma, also advocated by KNUDSON as a sentinel phenotype.

It would seem that, underlying the proposal of using a unique (or a few) indicator, is the hypothesis that all mutagens induce an increase in the probability of mutation at random over all loci, and that they are aspecific as far as the type of mutation is concerned. One would then expect that, if there is a significant increase in one indicator, this is correlated with a general increase.

4.2 Specificity of mutagens

For teratogens, it is well known, were it only from the thalidomide accident, that one teratogen will tend to cause the same type of malformation. The chemical variety of a developing enbryo may account for this specific behavior, in which there is only a specific chemical stage at a specific site which is prone to react with a teratogen. For mutagens, it is known that physical agents such as radiation may cause point and chromosome mutations, as well as aneuploidy and polyploidy (Tab. 2). However, the sequencies of chemical reactions leading to non-disjunction and polyploidy may be quite different from those leading to point and chromosome mutations. One might also attempt a classification of mutagens based on their specific (or aspecific) mutagenic effect, as proposed in the following table, in which an appropriate binary number is assigned to a given class of mutagen. The classification might be made as fine as desired, to include more detailed types of mutational events in different test organisms. Were most mutagens aspecific, it would suffice to test the induction of chromosome breaks in cultured cells to assess a product before possible delivery for use, and population monitoring would not have very high priority, since mutagens might be easily screened.

Table 2

Mutagen Type	Type of mutation induced				Examples of mutagens
	Polyploidy	Aneuploidy	Chromosomal	Point	
0	0	0	0	0	
1	0	0	0	1	Nitrous acid
2	0	0	1	0	Dyes
3	0	0	1	1	
4	0	1	0	0	Aging
5	0	1	0	1	
6	0	1	1	0	
7	0	1	1	1	
8	1	0	0	0	Colchicine
9	1	0	0	1	
10	1	0	1	0	
11	1	0	1	1	
12	1	1	01	0	
13	1	1	0	1	
14	1	1	1	0	
15	1	1	1	1	X-Ray

4.3 Detection of the mutagen

Were a mutagen specific, and capable of inducing a dominant effect, it might be detected a posteriori as thalidomide has been. The main danger, for human health, would appear to reside in aspecific mutagens which would individually escape screening and would reach the population. These would also seem the most difficult to detect a posteriori because of their aspecificity, particularly if "spy mutants" or "sentinal phenotypes" are the only appareils at the stations of a warning system. VOGEL (1970) wrote "... mutagenic noxes do not cause new syndromes; they merely increase non-specifically the probability of events which occur more rarely spontaneously anyway. Chances of being able to define an effect of this nature are virtually zero". Although some mutagens are specific, this statement reflects a situation which was true at the end of the sixties; it would seem now that some progress in monitoring is possible, and that the chances of detecting an effect are growing, and methodologies are developed which increase the chances well above zero. The basic idea would seem that, instead of using a single indicator, one may use all which are available.

5. The problem of monitoring and surveillance

5.1 Origin of the concern

Several systems have been proposed for screening chemical compounds having deleterious effects in man, be the effects carcino-, terato-, or mutagenic. For mutagenic effects, two-, three-, and multiple tier systems have been developed which use the mutagenic action of chemicals on bacteria, *Drosophila*, mice, and human cells in culture. No standardized system has been accepted, and this is possibly due to the absence of legislation for the a priori assessment of mutagenicity of chemicals in most countries.

Whatever the outcome of future government action toward the establishment of a priori screening, no system of screening will ever be completely efficient. Even disregarding the occurrence of accidental delivery of mutagens in the environment, which has happened often in the past, and will unavoidably happen again in the future, some mutagen will escape the series of filters which will be set up eventually, and for some time the population will be exposed to the action of such mutagens. This occurrence is the one which originates the concern for monitoring and surveillance of human populations for increased mutation.

5.2 Proposed systems of monitoring

The difficulties inherent in monitoring human populations for detecting the action of chemicals were recently examined by CROW (1971). For practical purposes, two main models have been proposed to date, one based on the concept of the "spy mutant" and on the use of a reduced fluctuation test for the revelation of significant deviations from a POISSON expectation (BARRAI, 1975); and the second based on the serial observation of a "sentinel phenotype" and on the application of sequential analysis over a POISSON expectation for the detection of significant deviations (MORTON and LINDSTEN, 1976). The use of the POISSON distribution for the expectation of mutant phenotypes, is based on the expectation of a constant number of such phenotypes, and derives immediately from the assumption that mutation rates are constant in space and time. The use of a POISSON expectation was also proposed for the monitoring of malformations and of teratogens (WEATHERALL et al., 1976); in principle, the methods described apply equally well to teratogens and to carcinogens beside to mutagens. The reduced fluctuation test was applied to data on genetic morbidity in British Columbia published by TRIMBLE and DOUGHTY (1974); it was sensitive enough to detect changes in ascertainment methods in that series.

The sequential analysis model was applied by MORTON and LINDSTEN on 16 consecutive years of births of Down's syndrome in Australian Victoria; for each year, they had two observations of incidence, one rural and the other urban. The sequential analysis was powerful enough to show a siginificant peak of incidence in 32 sequential observations (rural preceding urban incidence every year), and showed also that there was no consistent upward trend in the period of observation. The fluctuation test, applied to the same series, was highly significant ($\chi^2_{[31]} = 164.60$); however, when the series was broken in two series by rural and urban origin, fluctuation tests failed to show a significant yearly variation of the indicence. For the rural series the $\chi^2_{[15]}$ was 5.71, and for the urban series the $\chi^2_{[15]}$ was 11.87, none of them significant.

Sequential analysis has some advantage on the fluctuation test, in the sense that it requires less observations than a prospective study; also, it would detect the point in time at which the significant variation occurs. On the other hand, the fluctuation test shall detect the point in time when heterogeneity in a series becomes significant, this point not necessarily coinciding with the last observation.

A third method, which is proposed here, is based on the sequential study of the mean and variance of the cumulated incidences of available classificators of genetic morbidity available in one population. The sequential study of means and variances may be considered also for a single diagnostic group, should incidences in different parts of a region be available. In principle, the method is based in comparing means in subsequent years using the t-test, and in comparing variances in subsequent years using the F-test. The tests are then plotted on a chart which has the appropriate 5 per cent and 1 per cent limits as the quality control charts. An alert system is activated whenever a test reaches significance.

This method was applied to two different sets of data; however, before discussing the results, it is convenient to summarize some points which seem useful for the prospect of monitoring and surveillance.

5.3 The variation of mutation rates

It seems a safe assumption that mutation rates are constant in our time scale. Variation in space is reflected by estimates of the same mutation rate, which in average vary of a factor of three. Difference in rates at the same locus cannot be excluded, but it would seem safer to assume that they are due to the methodologies used for the collection and analysis of data and not to a real difference. The preceding statement indicates how difficult it would be to decide that a significant variation in mutation rates has taken place. If the available methods give estimates which can vary of a factor of three, where do we place the limit at which a significant increase is accepted? The use of a single entity, of a single phenotype may be quite unreliable at the present stage of development of the methodology. Unless the gene satisfies a set of conditions which are not easily met, it seems unlikely that a single entity is used as an indicator of increased mutation rates. Then, what can we use instead as an indicator of significant variation. It seems desirable to establish first a set of properties for a phenomenon useful to study the variation of genetic morbidity, which is due to fresh mutant genes only in part.

5.4 Properties of a phenomenon useful in the monitoring of genetic morbidity

The properties which seem more important are

a. Stability in time
b. Stability in space
c. Ease of observation and of data collection
d. Genetic or partially genetic etiology
e. Normally reported to some public service, such as health or vital statistics services
f. Frequency of the order of cumulative mutation rate, say of the order of one per cent.

Once the phenomenon is chosen for monitoring, on the basis that it seems to possess the required properties, an indicator has to be obtained from it, usually in the form of a statistical parameter capable of reflecting significant variation. The indicator will have some properties: it should be stable when the phenomenon is stable, it should be sensitive to the variation, and amenable to an exact test of differences.

6. Examples of applications

6.1 British Columbia data

The data on the amount of hereditary disease in British Columbia in 1952—1972 are an adequate set of data, permitting the application of the methods derived by quality control. The authors, TRIMBLE and DOUGHTY, classified the data in nine items, either true genetic entities, or partially genetic conditions, such as congenital malformations. In a previous analysis of these data, it was shown how it might be possible to detect significant variation in incidence in time, using a fluctuation test after correction of the absolute incidence to equal numbers of births. The data of TRIMBLE and DOUGHTY are collected as a special project, or as part of special projects; therefore, they do not satisfy the requirement of routine collection. The other requirements may be considered satisfied for the purposes of this analysis. The data refer to two distinct periods, each one characterized by a different ascertainment. The first period, which includes all births from 1952 to 1963, gives the data obtained from the British Columbia registry for handicapped children and adults. The second period, from 1964 to 1972, includes in addition records from the congenital anomalies surveillance system of British Columbia.

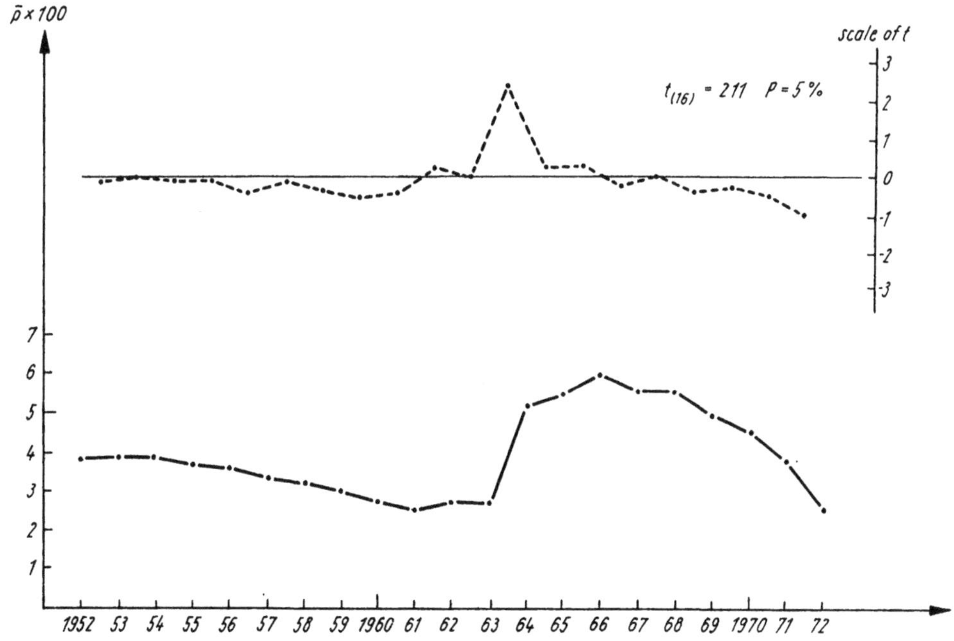

•——• CUMULATIVE INCIDENCE OF NINE ENTITIES HAVING COMPLETE OR PARTIAL GENETIC DETERMINATION. DATA FROM THE BRITISH COLUMBIA STUDY OF TRIMBLE AND DOUGHTY.

•----• t TEST BETWEEN ONE YEAR AND THE PRECEDING YEAR.

Fig. 1

There is a discontinuity between the two periods, in the sense that ascertainment becomes more complete in 1964, with a sudden jump in the number of cases per 100 liveborn; the average goes from about 3 per cent to about 5 per cent. This discontinuity in incidence, which is due to changed ascertainment, is most useful to us as an example, because it simulates a sudden increase in incidence, as might be due to a powerful non-specific mutagen and teratogen. In the following graph (Fig. 1), the average incidence is given. The sudden jump between 1963 and 1964 is conspicuous. Should all the data of this series have been collected under the same type of selection, the hypothesis of significant variation in the incidence due to some external factor would have emerged strongly from the simple observation of yearly means.

The data as collected by Trimble and Doughty are amenable also to the calculation of an even stronger indicator, which detects also variation in the individual incidences; this is the variance of the nine items classified in each year. In the next graph (Fig. 2), one may observe the behaviour of the variance in different years; the change from 1963 to 1964 is much more dramatic than the change in the means. Where the increase of the mean is about double, the variance increases five times.

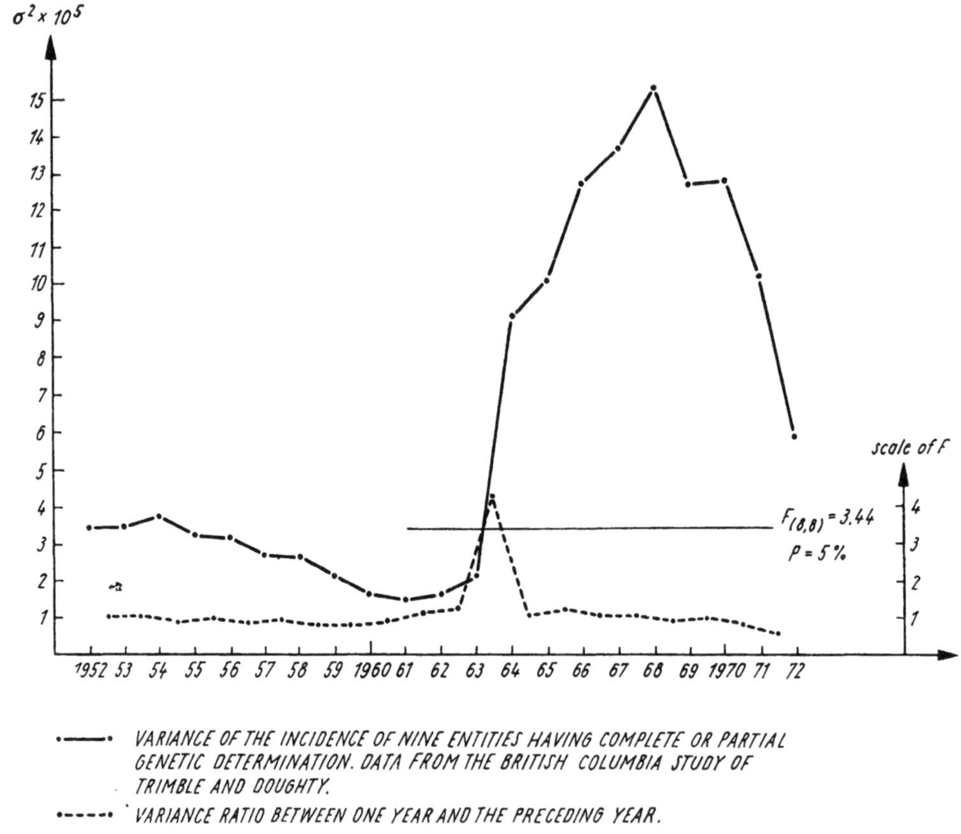

Fig. 2

Assuming then that it were possible for a health service to organize data collection in the same way as British Columbia, one might think of monitoring the significance of sudden changes in morbidity having hereditary origin, using the variance of the incidence of major groups instead of indicators extracted from a single entity or even a single group.

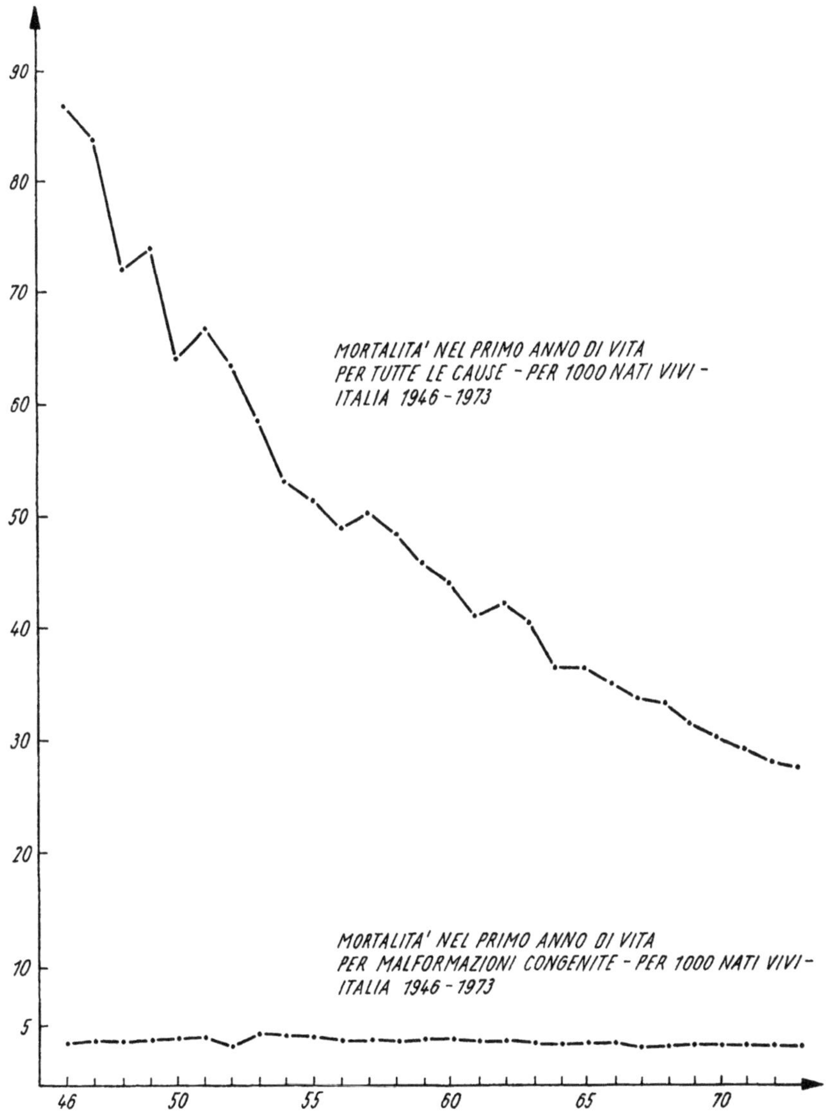

Fig. 3

6.2 Italy

In Italy, we have as yet no data comparable to those of British Columbia. However, in our quest for a phenomenon which would come close to satisfy the requirements which were considered necessary, we have studied in some detail infant mortality in Italy in the past 30 years. As it can be observed in the next graph, (Fig. 3), there has been a decrease of a factor of 3 in infant mortality since 1946. The decrease fits very well an exponential function.

The point which is of greater interest for our purpose, is the fraction of infant mortality which is due to congenital malformations. One may easily observe the remarkable stability of this phenomenon over the years; there is no obvious trend toward increase or decrease, and the mean stays close to 3.5 per thousand births per year. This phenomenon is stable in time, the data are easily collected and reported to the Italian Statistical Institute, and is obviously correlated with genetic factors. The frequency is about two orders of magnitude greater than individual mutation rates.

The Italian Statistical Institute publishes the incidense by each of the 20 regions of the country (Fig. 4) so that it was possible to compute the variance of the incidence in

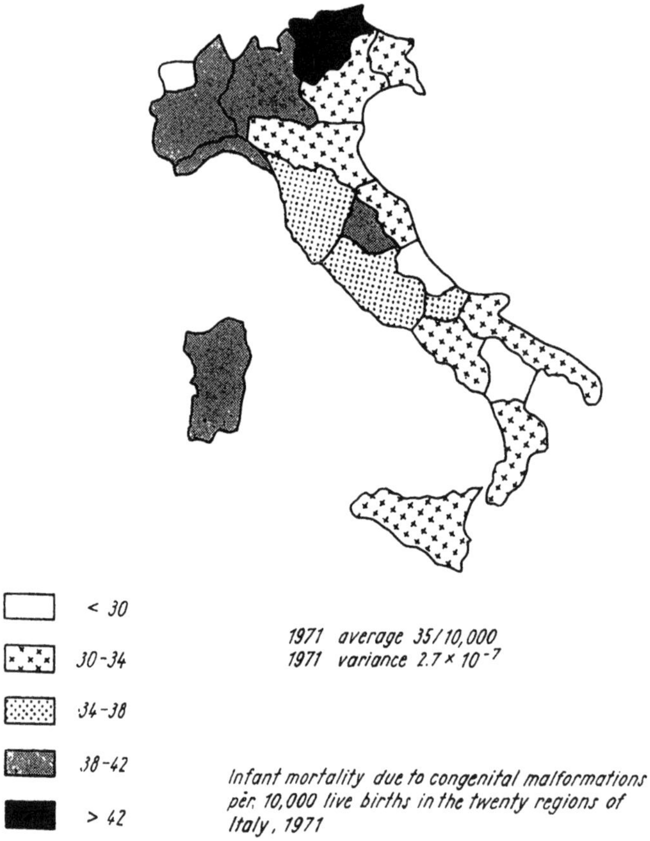

Fig. 4

regional values for the years available, 1968—1972. The means and variances for this period are given in the next graph (Fig. 5), and one may observe that the variance is not stable as the yearly mean for the whole country. There is no significant variation in the variance in consecutive years, although a trend toward decrease seems apparent. Data from previous years should be available to decide on the significance of the trend. However, even from these few years, it would be concluded that there is no factor producing strong regional variation of the incidence for the period considered.

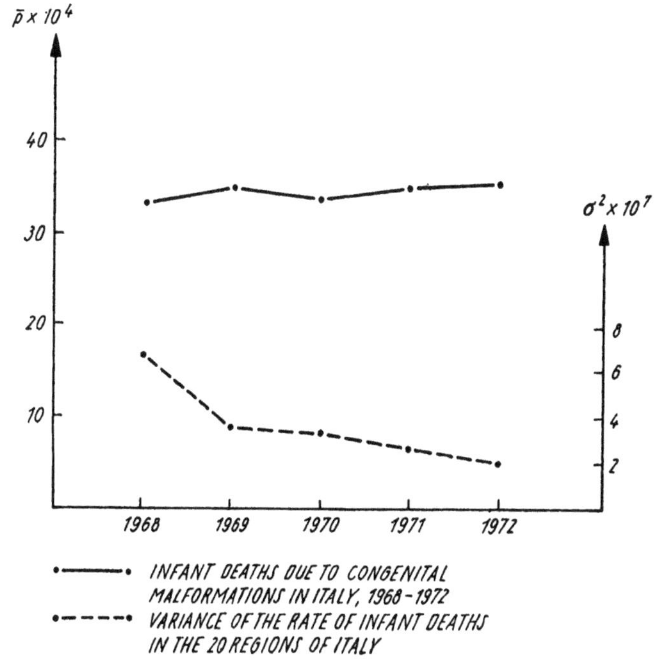

Fig. 5

The methods of comparing means and variance in consecutive years would not detect a trend toward increase or decrease; for such detection, one might however apply simple regression methods to the data.

7. Conclusions

In recent years, a few methods for the monitoring and surveillance of human populations have been developed. Former pessimism about monitoring for mutagens and teratogens should be reconsidered in the light of the available methodologies, and health statistical services might orient data collection to the application of the methods proposed.

References

[1] Barrai, I., Problems in the estimation of mutation rates. Mutation Research, **33** (1975) 37.

[2] Conneally, P. M., Mutation rates in man. Modern Trends in Human Genetics, **2** (1975) 204.

[3] Crow, J. F., Human populations monitoring. Chemical Mutagens, **2** (1971) 591.

[4] Dewey, W. J., I. Barrai, N. E. Morton and M. P. Mi, Recessive genes in severe mental defect. Am. J. Hum. Genet., **17** (1965) 237.

[5] Freire-Maia, A., Human genetic studies in areas of high natural radiation. VII. Genetic load. Rev. Bras. de Pesquisas Med. e Biol., **8** (1975) 287.

[6] Haldane, J. B. S., The rate of spontaneous mutation of a human gene. J. Genet., **31** (1935) 317.

[7] Jacobs, P. A., Epidemiology of chromosome abnormalities in man. The Cutter Lecture in Preventive Medicine given at the School of Public Health, Harvard University – May, 1976.

[8] Knudson, A. G. Jr., Mutation and cancer: statistical study of retinoblastoma. Proc. Nat. Acad. Sci. U.S.A., **68** (1971) 820.

[9] McKusick, V. A. (1971) Mendelian Inheritance in Man, third ed., Baltimore: Johns Hopkins Press

[10] Morton, N. E. and C. S. Chung, Formal genetics of muscular dystrophy. Am. J. Hum. Genet., **11** (1959) 360.

[11] Morton, N. E., J. F. Crow and H. J. Muller, An estimate of the mutational damage in man from data on consanguineous marriages. Proc. Nat. Acad. Sci. U.S.A., **42** (1956) 855.

[12] Morton, N. E. and J. Lindsten, (1976) Surveillance of Downs syndrome as a paradigm of population monitoring. Human Heredity, in press.

[13] Newcombe, H. B., Techniques for monitoring and assessing the significance of mutagenesis in human populations. Chemical Mutagens, **3** (1973) 57.

[14] Siniscalco, M. (1963) Discussion of genetic studies in isolates. In: Goldschmidt (1963), p. 203.

[15] Slatis, H., Seasonal variation in the American live birth sex ratio. Am. J. Hum. Genet., **5** (1953) 21.

[16] Stevenson, A. C., H. A. Johnston, M. I. P. Stewart and D. R. Golding, (1966) Congenital Malformations. Geneva: WHO

[17] Trimble, B. K. and J. H. Doughty, The amount of hereditary disease in human populations. Annals of Human Genetics, **38** (1974) 199.

[18] Vogel, F. (1970a) Spontaneous mutation in man. In: Chemical Mutagenesis in Mammals and Man, p. 16. Berlin: Springer

[19] Vogel, F. (1970b) Monitoring of human populations. In: Chemical Mutagenesis in Mammals and Man, p. 445. Berlin: Springer

[20] Vogel, F. and G. Röhrborn, (1970) Chemical Mutagenesis in Mammals and Man. Berlin: Springer.

[21] Weatherall, J. A. C. and J. C. Haskey, Surveillance of malformations. Brit. Med. Bull., 32 (1976) 39.

Occupational Hazards: Mutagenic Activity of Industrial Compounds

Nicola LOPRIENO

Laboratorio di Mutagenesi e Differenziamento, CNR, and Cattedra di Genetica, Università di Pisa.

1. Introduction

Today often emphasis is laid on linking modern technology to its negative effects on life conditions of large population and individuals, rather to the civil progress reached by modern man: the ideal value of technology, as a direct expression of man's culture, is no the hand ignored.

The history of technology represents an important aspect of man's culture, in his continuous effort to take possession of naturral resources.

In the general scheme of modern technology, the present development of chemical production has produced a large number of potential negative effects on man. A knowledge and an understanding of this is still in its initial stage and has many difficulties, if one realises that every year between 500 and 1000 new chemical compounds are put on the market; the chemical industry is developing at an average rate of 10% per year. The attempts to protect the worker's health and the general public environment against the deleterious effects of chemical compounds have been difficult, because of little interest of the present industrial system on these particular problems, and of paucity of financial resources devoted to the solution of these problems. Another source of difficulties has arisen directly from the limitations existing in the development of reliable toxicological procedures which, till now, have been unable to predict a series of negative effects of chemicals for man.

Among biological long term effects of those chemicals reactive with basic cellular molecules or subjected to a very active mammalian metabolism, mutagenic effects displayed by substances present in working environment are largely not yet evaluated, because only recently the list of chemicals which have been submitted to mutagenic tests has been increased: for many years mutation research on chemicals has been mainly interested in the study of basic molecular mechanism of mutational process. Only when it was established that some chemical compounds were found strong mutagens for several laboratory biological systems and that they were representative of several classes of molecules presented in many chemical processes, did their potential genetic hazard for the population stimulate some interest. Due to a strong development of research in the field of environmental mutagenesis in several countries, the relevance of mutagenic analyses and the need of an assessement of the potential mutagenic activity of many compounds present in our environment have been stressed everywhere, and a large interest in these new toxicological approaches have been expressed by general public, labour trade unions and industrial organizations.

Among employees organizations, during the last Conference of European Trade Unions on the Humanization of the Work Environment [1] it has been specifically requested an improvement of the health pretection against the dangerous effects of a professional exposure to toxic compounds which might cause tumours and "genetic alterations". The Commission of the European Communities is discussing during this year a proposal for a new regulation of definition, labelling and marketing of dangerous substances produced by chemical industries [2]: among other toxicological properties, this proposal is requiring an evaluation of the "mutagenic" activity of the chemicals.

2. Evaluation of chromosomal mutations

One of the most widely used methodology to assess the potential mutagenic activity of a compound present in worker's environment consists of the analysis of chromosome aberrations in somatic cells (peripheral blood lymphocytes) of subjects exposed to the chemicals.

The production of chromosome aberrations has mainly been used as a biological indicator of a potential mutagenic effect of professional exposure, but it has since now never been correlated with epidemiological studies showing a possible occurrence of genetic alterations among the progeny of the exposed individuals or other genetic deleterious effects. Examples of an attempt to correlate chromosome aberrations in lymphocytes with other pathological symptoms have been presented in some recent papers [3, 4, 5, 6].

This methodology is so far the only mutagenic evaluation that may be applied in vivo directly on employees exposed to compounds with mutagenic activity presents in chemical factories. Table 1 indicates the major classes of chemicals, to which an occupational exposure is known, that have been found positive for their clastogenic activity [7]: these data have been derived by a direct monitoring of peripheral blood

Table 1

Classes of chemical agents known to induce chromosome changes in human cells

Classes of compounds	Type of exposure
1. Alkylating agents	occupational
2. Haloethers	occupational
3. Solvents	occupational
4. Inorganic chemicals	occupational, environmental
5. Pesticide, Fungicides, Herbicides	occupational, environmental
6. Food additive	environmental
7. Air and Water Pollutants	environmental

cells from exposed workers, as well from analyses after in vitro treatments of human lymphocytes or fibroblasts on variyng times and concentrations.

It is well established that such an analysis does not represent a suitable methodology for assessing the mutagenicity in vitro or in vivo of a given chemical or for defining the genetic hazard for men exposed to chemicals, as the extrapolation of the results to germ cells remains disputable [8], although it is well established that among human genetic disorders, chromosome aberrations represent an incidence of 0.5—1.0% [9].

3. Laboratory mutagenicity tests

An analysis of the mutagenic activity of chemicals of occupational hazard should taken into consideration a high number of compounds that are the result of the chemical industry production or an even higher number of compounds that are employed in the industrial process. In all cases, such a number is so high, whereas a very small number of biological data of different types have been up to now collected, including those on mutagenicity. The best estimate of the number of chemicals so far analyzed for their mutagenic activity is around 1000 [10], whereas the number of compounds in significant production of some interest for the occupational mutagenic hazard could be estimated of about 25000 from data available in United States. Table 2 reports classes and types of chemical molecules submitted to mutagenic analyses.

Table 2

List of chemicals analyzed for their mutagenic and clastogenic activities

1. Alkylating agents: Aziridines (34), Triazines (20), Nitrogen, Sulfur, and Oxide mustards (56), Phosphoric acid esters (21), Epoxides (15), Lactonyes (5), Aldehydes (10), Alkylsulfates (5), Alkane sulfonic esters (15), Diazoalkanes (6), Aryldialkyltriazenes (12), Alkyl and alkane halides (47), Azoxy and hydrazoalkanes (4).
2. N-nitroso compounds: Nitrosamines (22), Nitrosamides (10).
3. Organic peroxides: (5).
4. Polynuclear aromatics: (23).
5. Heterocycles: Benzimidazoles (21), Fluorenones (16), Phenothiazines (7).
6. Natural products: Mycotoxins (13), Pyrrolizidine alkaloids (8), Antibiotics (101), Xanthines (9), Steroids (9).
7. Halogenated ethers and halohydrins: (8).
8. Halogenated hydrocarbons and related derivatives: Vinyl and vinylidiene derivatives (13), Halogenated aromatics (55), Flurocarbons (10).
9. Nucleic acid bases and analogs: (89).
10. Hydrazines, hydroxylamines, carbamates, hydrazides, and ureas: Hydrazines (5), Hydroxylamines (20), Carbamates (24), Hydrazides (6), Ureas and thioureas (28).
11. Aromatic amines: (42).
12. Azo dyes: (26).
13. Nitro derivatives: Nitroquinolines and related compounds (4), Nitrofurans (49), Nitroimidazoles (5).
14. Organo-metallics: Organo lead derivatives (5), Organo mercury derivatives (8), Others (32).
15. Miscellaneous: Ester and anhydrides (40), Quaternary Ammonium (9), Quinones (23), N-oxides (14), Sulfites (1), Sultones (2), Cyclodienes (7), Nitrosoureas (22).

A second problem is related to the number of employees: in order to consider a given chemical for its potential mutagenic occupational hazard, there must be a considerable number of exposed people at risk. Data presenting figures on the number of workers employement are difficult to obtain, whereas the amount of production of the most important chemicals is relatively well known. For instance, the data presented in Table 3 are of great value for understanding this problem: examples are presented for compounds of known mutagenic activity.

On the basis of the total yearly production, 6 chemicals have been selected, which are reported in Table 4, found among those of highest production during 1975.

Table 3

Compounds of known mutagenic activity, their productions and number of employees in USA

Compounds	Production $\times 10^9$ Kg/year	No. of workers employed
Vinyl chloride monomer	2.5	1 500
Trichloroethylene	0.2	283 000
Chloroprene	0.2	2 500
Vinylidiene chloride	0.1	100

Table 4

Chemicals of relevant production during 1975 in USA[1]

Compounds	Formula	Production $\times 10^9$ lbs/year
Ethylene dichloride (16)	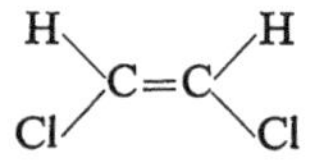	5.94
Formaldehyde (22)	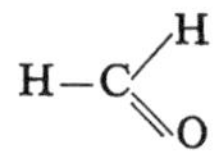	4.62
Styrene (23)	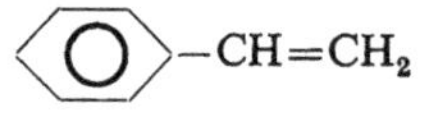	4.40
Vinyl chloride (24)	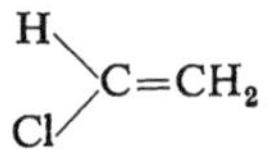	4.20
Ethylene oxide (26)	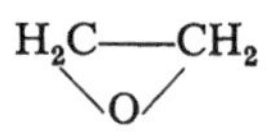	3.94
Propylene oxide (43)	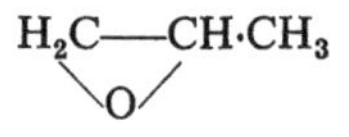	1.52

[1] Data from Chem. & Eng. News May 3, 1976, pag. 33–48

[2] Rank no. during 1975.

3.1. Ethylene dichloride has been employed in many different industries of agricultural interest and in some industrial chemical process, such as the production of vinyl chloride monomer, trichloroethylene, etc. It has been found mutagenic in *Salmonella*, without metabolic activation [11, 12, 13] and able to inhibit the growth of *pol A^-* mutant of *E. coli* [14]. Moreover it is known to produce recessive lethals [15], chromosome non-disjunction [16], and X-linked recessive mutations [16] in *Drosophila*.

3.2. Formaldehyde, a widely distributed compound in man's environment, is employed mainly in the synthetic resin industry and in the manifacturing industries. It has been found mutagenic for *E. coli* [17], *Neurospora* [18] and *Drosophila* [19]; it induces, moreover, recombination and gene-conversion in *S. cerevisiae* [20, 21] and in *Drosophila* [22].

3.3. Styrene is used mainly in the production of plastics, resins or rubber, in association with other polymers, such as acrylonitrile or butadiene. Miscellaneous applications of styrene-oxide, an intermediate in the preparation of different chemicals of agricultural or biological interest. Styrene has been found mutagenic after metabolic activation on *Salmonella*, in vitro [23], or in the host mediated assay on *Salmonella* [13] or yeast [24]; moreover it has to be considered an indirect mutagen, as its mammalian metabolic conversion production, styrene-epoxide, has been found mutagenic on several indicator organisms, such as *Salmonella* [25], yeast [24], hamster cells (azaguanine-resistant mutants) [24]; it induces unscheduled DNA synthesis in human cell line [26].

3.4. Vinyl chloride, whose carcinogenic activity towards man, as well on animals, has been well established [27], is a compound of large use mainly in the preparation of plastic material employed in different fields. Its mutagenic activity has been demonstrated after metabolic activation on *Salmonella* [28], yeasts [29], *Drosophila* [30]; it produces chromosome aberrations in human lymphocytes, in vivo [31], and point mutations on yeast, in the host mediated assay [29]. Many of its in vivo metabolites have been studied as mutagens in different biological systems [32].

3.5. Ethylene oxide has been used as a fungicide for treatment by fumigation of many different types of material, including foods; it is used as an intermediate in the industrial production of ethylene-glycols and other compounds [33]. It has been found to react with DNA, by alkylating the N-7 position of guanosine [34]. It is known to produce mutations in *Salmonella* [35], *Neurospora* [36]; it induces recessive lethals [37], translocations [38] and minute mutations [39] in *Drosophila*. Dominant lethal mutations and chromosome aberrations in bone-marrow cells in rats were also induced [40].

3.6. Propylene oxide is used as an intermediate in the production of polyether polyols, that are used for preparing polyurethane foams; it is employed also in the production of propylene glycol and in the preparation of unsaturated polyester resins.

It is known as an alkylating agent for both guanosine and adenosine [41]: it is mutagenic for yeast [42], *Neurospora* [43] and produces recessive lethals in *Drosophila* [37].

Moreover, compounds of large production to which a large fraction of workers is exposed, such as vinylidiene chloride, trichloroethylene, carbon tetrachloride, chloroform, tetrachloroethylene, haloethers have so far receive very little attention and the majority of them has been tested only on *Salmonella*, and some of them have been found mutagenic after metabolic activation.

4. Mutagenic analyses of urine of exposed workers

Since some metabolite of numerous chemical substances with high biological reactivity are produced in vivo in the mammalian body and they are excreted with urine, it would be possible by monitoring the urines with mutagenic analyses to reveal the presence of metabolites and their concentration in the human urine for each individual worker which has been exposed for some time to an environmental direct or indirect mutagen: several indicator organisms may be used for a mutagenic analysis of the urine, and *Salmonella* and yeasts have already been applied. On analyzing urine from individuals in the population for determining whether or not active metabolites of presumptive mutagens detected in the environment are present, would indicate that the individual worker is at risk from the substance. Up to now such a methodological in vivo monitoring has been develop for human patients submitted to treatments with different drugs [44] and not yet used in the problem of worker's health, with the exception of a recent communication in which the urines of workers exposed to vinyl chloride monomer in a factory have been tested for their possible mutagenic activity [45].

5. Conclusions

The principal comments that may be expressed on the data available so far on the mutagenic activity of such an important industrial and commercial class of compounds is as follows: with the exception of some very few that have been tested also in *Drosophila*, or with hamster cells (vinyl chloride and styrene, or their metabolites), the majority of the compounds which have already been shown to produce cancer in test animals or man, have been analyzed for their mutagenic effects only on microbiol systems, such as *Salmonella*, *Neurospora*, and yeasts. Mammalian metabolic systems have been employed in the mutagenic analyis, stressing therefore the importance of the metabolic conversion in the expression of mutagenic activity of these different compounds (styrene, vinyl chloride, vinylidiene chloride, trichloroethylene, chloroprene). They have demonstrated its potential mutagenic risk for mammalian organisms, but it is not possible to indicate the genetic risk they eventually represent for the population in general and for directly exposed workers in particular.

There is a great demand from Labour Trade Unions in different countries at the present, to develop all possible technical and scientific information that are needed for making a decision about the existence of genetic risk for workers professionally exposed in a chronic way to many of these compounds. Up to now the professional exposure is protected only on the basis of the Maximum Allowable Concentration (MAC) which represents a safe condition only for those effects clinically determined.

References

[1] Conference of European Trade Union Centres, Genève, 28 February—1 March 1975, The protection of workpeople from the effects of dangerous substances.

[2] Project de proposition de modification de la directive du Conseil du 27 Juin 1967 concernant le repprochement des dispositions legislatives, réglementaires et administratives relatives aux substances dangereuse (67/548)CEE). Commission Des Communautes Européennes.

[3] Czeizel, A., I. Szentesi, E. Hornyak, Gy. Ungvari, Z. Bognar, M. Timar, Genetic study on PVC workers. Sixth Annual Meeting of EEMS, 81, 1976 (Abs.).
[4] Kiraly, J., A. Czeizel, I. Szentesi, Genetic study on workers producing organophosphate insecticides. Sixth Annual Meeting of EEMS, 109, 1976 (Abs.).
[5] Kucerova, M., V. S. Zhurkov, Population cytogenetic study of mutagenic effect of Epichlorohydrin. Sixth Annual Meeting of EEMS, 119–120, 1976 (Abs.).
[6] Zudova, Z., K. Landa, Genetic risk of occupational exposure to haloethers. Sixth Annual Meeting of EEMS, 165, 1976 (Abs.).
[7] Shaw, M. W., Human chromosome damage by chemical agents. Ann. Rev. Med., **21** (1970) 409–432.
[8] Leonard, A., Heritable chromosome aberrations in mammals after exposure to chemicals. Rad. and Environm. Biophys., **13** (1976) 1–8.
[9] Nielsen, J., I. Sillesen, Incidence of chromosome aberrations among 11, 148 newborn children. Humangenetik, **30** (1975) 1–12.
[10] Draft Document on Methods for Determining the Mutagenic Properties of Chemicals: DHEW Committee, January 26, 1976.
[11] McCann, J., V. Simmon, D. Streitweiser, and B. N. Ames, Mutagenicity of chloroacetaldehyde, a possible metabolic product of 1,2-dichloroethana (ethylene dichloride), chloroethanol, and cyclophosphamide (ethylene chlorohydrin), vinyl chloride. Proc. Natl. Acad. Sci. (US), **72** (1975) 3190–3193.
[12] Bignami, M., G. Cardamone, P. Comba, V. A. Ortali, G. Morpurgo, and A. Carere, Relationship between chemical structure and mutagenic activity in some pesticides: the use of *Salmonella typhimurium* and *Aspergillus nidulans*. Sixth Annual Meeting of EEMS, 67, 1976 (Abs.).
[13] Cerna, M. and H. Kypenova, Mutagenic activity of chloroethylenes analyzed by screening system tests. Sixth Annual Meeting of EEMS, 77, 1976 (Abs.).
[14] Brem, H., A. B. Stein and H. S. Rosenkranz, The mutagenicity and DNA – modifying effect of haloalkanes. Cancer Res., **34** (1974) 2576–2579.
[15] Rapoport, I. A., Reaction of gene proteins with ethylene dichloride. Akad. Nauk. S.S.S.R. Dokl. Biol. Sci., **134** (1960) 745–747.
[16] Shakarnis, V. F., 1,2-dichloroethane-induced chromosome non-disjunction and recessive sex-linked lethal mutation in *Drosophila melanogaster*. Genetika, **5** (1969) 89–95.
[17] Nishioka, H., Lethal and mutagenic action of formaldehyde in Hcr^+ and Hcr^- strains of *Escherichia coli*. Mutation Res., **17** (1973) 261–265.
[18] Dickey, F. H., G. H. Cleland and C. Lotz, Proc. Natl. Acad. Sci. (US), **35** (1949) 581.
[19] Nafei, H. and C. Auerbach, Mutagenesis by formaldehyde food in relation to DNA replication in *Drosophila* spermatocytes. Z. Vererbungsl., **95** (1964) 351–367.
[20] Chanet, R., C. Izard, and E. Moustacchi, Genetic effects of formaldehyde in yeast. I. Influence of the growth stages on killing and recombination. Mutation Res., **33** (1975) 179–186.
[21] Chanet, R., C. Izard, and E. Moustacchi, Genetic affects of Formaldehyde in yeast. II. Influence of ploidy and of mutations affecting radiosensitivity on its lethal effect. Mutation Res., **35** (1976) 29–38.
[22] Alderson, T., Induction of genetically recombinant chromosomes in the absence of induced mutations. Nature, **215** (1967) 1281–1283.
[23] Vainio, H., R. Pääkkönen, K. Rönnholm, V. Raunio, O. Pelkonen, A study on the mutagenic activity of styrene and styrene oxide (in press).
[24] Loprieno, N., A. Abbondandolo, R. Barale, S. Baroncelli, S. Bonatti, G. Bronzetti, A. Cammellini, C. Corsi, G. Corti, D. Frezza, C. Leporini, A. Mazzaccaro, R. Nieri, D. Rosellini, A. M. Rossi, Mutagenicity of industrial compounds: Styrene and its possible metabolite styrene-oxide. Mutation Res. 1976 (in press).
[25] Milvy, P. and A. J. Garro, Mutagenic activity of styrene oxide (1,2-epoxyethyl-benzene), a presumed styrene metabolite. Mutation Res., **40** (1976) 15–18.
[26] Abbondandolo, A., S. Bonatti, G. Corti, R. Fiorio, and A. Mazzaccaro, Microsome-mediated assay with cultured mammalian cells. Sixth Annual Meeting of EEMS, 51, 1976 (Abs.).

[27] BARTSCH, H. and R. MONTESANO, Mutagenic and carcinogenic effects of vinyl chloride. Mutation Res., **32** (1975) 93–114.
[28] BARTSCH, H., C. MALAVEILLE, and R. MONTESANO, Human, rat and mouse liver-mediated mutagenicity of vinyl chloride in *S. typhimurium* strains. Int. J. Cancer, **15** (1975) 429–437.
[29] LOPRIENO, N., R. BARALE, S. BARONCELLI, C. BAUER G. BRONZETTI, A. CAMMELLINI, G. CERCIGNANI, C. CORSI, G. GERVASI, C. LEPORINI, R. NIERI, A. M. ROSSI, G. STRETTI, G. TURCHI, Evaluation of the genetic effects induced by vinyl chloride monomer (VCM) under mammalian metabolic activation: Studies in vitro and in vivo. Mutation Res., **40** (1976) 85–96.
[30] VOGEL, E., The relation between mutational pattern and concentration by chemical mutagens in *Drosophila.* In R. MONTESANO, H. BARTSCH and L. TOMATIS: Screening tests in chemical carcinogenesis, IARC Scient. Publ., No. 12 (1976) 117–137.
[31] FUNES-CARVIOTO, F., B. LAMBERT, J. LINDSTEN, L. EHRENBERG, A. T. NATARAJAN, S. OSTERMAN-GOLKAR: Chromosome aberrations in workers exposed to vinyl chloride. Lancet 1, 459, 1975.
[32] BARTSCH, H. and N. LOPRIENO: Vinyl chloride (VCM): an example for evaluating adverse biological effects in shortterm tests. Sixth Annual Meeting of EEMS, 33–34, 1976 (Abs.).
[33] IARC Monograph on the Evaluation of Carcinogenic Risk of Chemicals to Man: Cadmium, nickel, some epoxides, miscellaneous industrial chemicals and general considerations on volatile anesthetics. **11** (1976) 157–167.
[34] BROOKES, P. & P. D. LAWLEY, The alkylation of guanosine and guanylic acid. J. Chem. Soc., (1961) 3923–3928.
[35] RANNUG, U., R. GÖTHE and C. A. WACHTMEISTER, The mutagenicity of chloroethylene-oxide, chloroacetaldehyde, 2-chloroethanol and chloroacetic acid, conceivable metabolites of vinyl chloride. Chem.-Biol. Interact., **12** (1976) 251–263.
[36] KØLMARK, G. and M. WESTERGAARD, Further studies on chemically induced reversions at the adenine locus of Neurospora. Hereditas, **39** (1953) 209–224.
[37] RAPOPORT, I. A., Action of ethylene oxide, glycidol and glycols on gene mutation. Dokl. Akad. Nauk. SSR, **60** (1948) 469–472.
[38] NAKAO, Y. and C. AUERBACH, Test of a possible correlation between cross-linking and chromosome breaking abilities of chemical mutagens. Z. Vererbungsl., **92** (1961) 457–461.
[39] FAHMY, O. G. and M. J. FAHMY, Gene elimination in carcinogenesis: re-interpretation of the somatic mutation theory. Cancer Res., **30** (1970) 195–205.
[40] EMBREE, J. W. and C. H. HINE, Mutagenicity of ethylene oxide. Toxicol appl. Pharmacol., **33** (1975) 172–173.
[41] LAWLEY, P. D. and M. JARMAN, Alkylation by propylene oxide of deoxyribonucleic acid, adenine, guanosine and deoxyguanylic acid. Biochem. J. **126** (1972) 893–900.
[42] HESLOT, H., Étude quantitative de réversions biochimiques induites chez la Levure *Schizosaccharomyces pombe* par des radiations et des substances radiomimetiques. Abhandlungen der Deutschen Akademie der Wissenschaften zu Berlin, 193–228, 1962.
[43] KØLMARK, G. and N. H. GILES, Comparative studies of non epoxides as inducers of reverse mutations in *Neurospora.* Genetics, **40** (1955) 890–902.
[44] SIEBERT, D. and U. SIMON, Genetic activity of metabolites in the ascitic fluid and in the urine of a human patient treated with cyclophosphamide. Induction of mitotic gene conversion in *Saccharomyces cerevisiae.* Mutation Res., **21** (1973) 257–262.
[45] MATTERN, I. E. and W. B. VAN DER ZWAAN, Mutagenicity testing of urine from vinyl chloride (VCM) treated rats using the *Salmonella* test system. Sixth Annual Meeting of EEMS, 127–128, 1976 (Abs.).

Vinyl Chloride (VCM): An Example for Evaluating Adverse Biological Effects in Short-Term Tests

Helmut BARTSCH and Nicola LOPRIENO

International Agency for Research on Cancer,
150 cours Albert Thomas, 69008 Lyon, France;
Laboratorio di Mutagenesi e Differenziamento,
CNR, Cisanello 147, 56100 Pisa, Italy

The various adverse biological effects of vinyl chloride (VCM) for review, see BARTSCH and MONTESANO, 1975 [1], a recognized carcinogen in animals [2, 3, 4, 5, 6] and man [7, 8, 9, 10], appear to be largely dependent upon its metabolic conversion into chemically reactive metabolites [1]. Various genetically transmissable or other biological changes of genetic interest caused by VCM or its metabolic products, have been observed, including cell lethality [11], reverse and forward mutations in bacterial [12, 13, 14, 15, 16, 17], yeast [18], or mammalian cells [19], recessive lethal mutations in *Drosophila melanogaster* [20], gene conversion in yeast [18, 21] and chromosome aberrations in human lymphocytes from peripheral blood [22, 23, 24].

As the liver-microsome mediated formation of two chemically reactive products, chloroethylene oxide [25] and 2-chloroacetaldehyde [26] in vitro from VCM has been reported, the biological activity of the latter two compounds, of monochloroacetic acid, an identified urinary metabolite in rats and man [27], and a putative metabolic intermediate, chloroethanol [28], were assayed for the detection of carcinogens in a collaborative study in several prescreening tests:

(a) assay for electrophilic reactivity with 4-(4-nitrobenzyl)-pyridine [25]
(b) tissue-mediated mutagenicity tests with *S. typhimurium* [12, 13, 29]
(c) tissue-mediated mutagenicity assay with *S. pombe* and *S. cerevisiae* [18, 21]
(d) in vitro induction of oubain and 8-azaguanine resistant mutations in Chinese hamster V 79 cells [19]
(e) induction of DNA repair synthesis, measured by unscheduled incorporation of ^{3}H-thymidine in cultured human skin fibroblasts [30] and
(f) host-mediated assay in mice with *S. pombe* [21]

Chloroethylene oxide was the most active compound in the assay systems (a) to (e), while 2-chloroacetaldehyde induced histidine reversion in *S. typhimurium* strains [13, 16, 17, 31], 8-azaguanine resistance in Chinese hamster cells [19], but was inactive in all other aforementioned assays. By comparing chloroethylene oxide with the less reactive 2-chloroacetaldehyde, quantitative differences in chemical reactivity by a factor 10^3 [25], or in mutagenic potency in *S. typhimurium* or *E. coli* by a factor 10^3 or 10^4, respectively have been reported [15, 17]. 2-Chloroethanol either showed no or marginally to weak activity in *S. typhimurium*, *S. pombe* and in the induction of DNA repair synthesis [13, 17, 21 and San & Stich, unpublished data]. No biological activity could be detected with monochloroacetic acid, a compound which was highly toxic.

The sequence of sensitivity to detect chloroethylene oxide in the various short-term tests mentioned above was in the following order (in brackets, estimated lowest detectable concentration in μmol/l): histidine reversion in *S. typhimurium* TA 1530 or TA 1535 (100); gene conversion in *S. cerevisiae* (10); unscheduled incorporation of ^{3}H-thymidine in cultured fibroblasts (20); forward mutation in *S. pombe* (10); induction of 8-azaguanine resistance in Chinese hamster cells (6) and colour reaction with 4-(4-nitrobenzyl)-pyridine (2).

Chloroethylene oxide induced local tumours in mice upon repeated subcutaneous administration with a frequency comparable to bis(chloromethyl)ether, a recognized human and animal carcinogen [32] when both compounds were tested at maximal tolerated toxic doses (ZADJELA, unpublished data).

Further evidence that hepatic mixed function oxidase is also involved in the activation of VCM in human liver, a target organ in carcinogenesis, is given by a plot of benzo-(a)pyrene hydroxylase (AHH) activity in 5 liver specimens, versus their capacity to metabolize VCM into mutagens (Fig. 1). The mutagenic activity is expressed as the number of histidine revertant colonies of *S. typhimurium* TA 1530 strain per 38 mg of wet tissue per plate. The AHH activity is given on the ordinate as pmol of 3-hydroxy-benzo(a)pyrene formed per minute per gramme of wet tissue. Values are taken from the linear region of time- or dose-dependent assays.

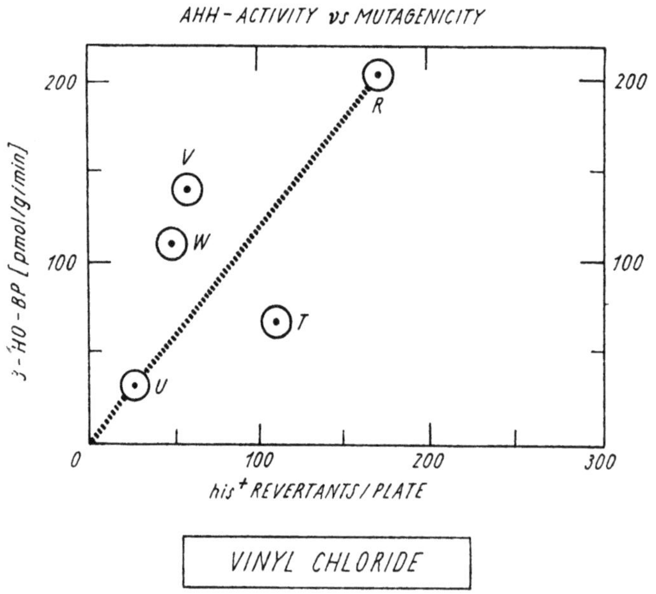

Fig. 1. AHH activity and tissue-mediated mutagenicity of vinyl chloride in human liver biopsies

Even from this limited number of samples, a positive correlation between the two enzymic activities can be deduced. The data are the first to demonstrate that microsomal benzo(a)pyrene hydroxylase in human liver activates VCM into electrophilic and mutagenic reactants.

These results strongly support the hypothesis that chloroethylene oxide is one of the principal mutagenic and/or carcinogenic intermediates formed from VCM by mammalian metabolism.

Activities are expressed as either pmoles of 3-hydroxy-benzo(a)pyrene formed per minute per g of wet tissue (ordinate) [33] or as the number of histidine-revertant colonies of *S. typhimurium* TA 1530 strain after 4 hours of exposure to 20% VCM in air per 38 mg of wet tissue per plate [12]. Different human liver biopsies are represented by different letters. Spontaneous mutations are subtracted; values are taken from linear dose- and time-dependent response curves.

Acknowledgements

Financial support for one of the author's (H. Bartsch) research activities in this field was partielly provided by contract No. 1CP-55630 of the National Cancer Institute, NIH (USA). The human tissues were kindly provided by Drs. M. Boiocchi, G. Della Porta and U. Veronesi, Istituto Nazionale per lo Studio e la Cura del Tumori, Milan, Italy. The authors are grateful to Drs. San and Stich, Vancouver (Canada), Dr. Zadjela, Orsay (France), Drs. Verburgt and Vogel, Leiden (Netherlands), for providing data prior to publication.

For valuable discussions and/or experimental assistance, the authors are indebted to Drs. L. Tomatis, R. Montesano, C. Malaveille, N. Sabadie, A. Barbin, A. Camus, G. Planche, G. Brun and H. Brésil.

We thank Miss P. Stafford Smith and Mrs E. Ward for their help in preparing the manuscript and for editorial aid.

References

[1] Bartsch, H. and R. Montesano, Mutagenic and carcinogenic effects of vinyl chloride, Mutation Res., **32** (1976) 93–114.

[2] Viola, P. L., Carcinogenic effect of vinyl chloride, Abstract X Int. Cancer Congress, Houston, Vol. 29 (1970).

[3] Viola, P. L., A. Bigotti and A. Caputo, Oncogenic response of rat skin, lungs and bones to vinyl chloride, Cancer Res., **31** (1971) 516–522.

[4] Maltoni, C. and G. Lefemine, Carcinogenicity bioassays of vinyl chloride, Environm. Res., **7** (1974) 387–405.

[5] Maltoni, C., G. Lefemine, P. Chieco and D. Carretti, La cancerogenesi ambientale e professionale. Nuove prospettive alla luce della cancerogenesi da cloruro di vinile, Gli Ospedali della vita, **1**, 5/6 (1964) 7–66.

[6] Caputo, A., P. L. Viola and A. Bigotti, Oncogenicity of vinyl chloride at low concentrations in rats and rabbits, IRCS, **2** (1974) 1582.

[7] Creech, J. L. and M. N. Johnson, Angiosarcoma of the liver in the manufacture of polyvinyl chloride, J. Occup. Med., **16** (1974) 150–151.

[8] Heath, C. W., H. Falk and J. L. Vreech, Characteristic cases of angiosarcoma of the liver among vinyl chloride workers in the United States, Ann. N.Y. Acad. Sci., **246** (1975) 231–236.

[9] Lee, F. I. and D. S. Harry, Angiosarcoma of the liver in a vinyl chloride worker, Lancet, 1 (1974) 1316–1318.

[10] IARC Monographs on the Evaluation of Carcinogenic Risk of Chemicals to Man, Some antithyroid and related substances, nitrofurans and industrial chemicals, International Agency for Research on Cancer, Vol. **7**, 1974, Lyon.

[11] Rosenkranz, S., H. S. Carr and H. S. Rosenkranz, 2-Haloethanols: mutagenicity and reactivity with DNA, Mutation Res., **26** (1974) 367–370.
[12] Bartsch, H., Malaveille, C. and R. Montesano, Human, rat and mouse liver-mediated mutagenicity of vinyl chloride in *S. typhimurium* strains, Int. J. Cancer, **15** (1975) 429–437.
[13] Malaveille, C., H. Bartsch, R. Montesano, A. Barbin, A.-M. Camus, A. Croisy and P. Jacquignon, Mutagenicity of vinyl chloride, chloroethylene oxide, chloroacetaldehyde and chloroethanol, Biochem. Biophys. Res. Commun., **63** (1975) 363–370.
[14] Greim, H., G. Bonse, Z. Radwan, D. Reichert and D. Henschler, Mutagenicity in vitro and potential carcinogenicity of chlorinated ethylenes as a function of metabolic oxirane formation, Biochem. Pharmacol., **24** (1975) 2013–2017.
[15] Hussain, S. and S. Osterman-Golkar, Comment on the mutagenic effectiveness of vinyl chloride metabolites, Chem. Biol. Interactions, **12** (1976) 265–267.
[16] McCann, J., V. Simon, D. Streitweiser and B. N. Ames, Mutagenicity of chloroacetaldehyde, a possible metabolic product of 1,2-dichloroethane (ethylene dichloride), chloroethanol and cyclophosphamide (ethylene chlorohydrin), vinyl chloride, Proc. Nat. Acad. Sci. (US), **72** (1975) 3190–3193.
[17] Rannug, U., R. Göthe and C. A. Wachtmeister, The mutagenicity of chloroethylene oxide, chloroacetaldehyde, 2-chloroethanol and chloroacetic acid, conceivable metabolites of vinyl chloride, Chem.-Biol. Interactions, **12** (1976) 251–263.
[18] Loprieno, N., R. Barale, S. Baroncelli, C, Bauer, G. Bronzetti, A. Cammellini, G. Cercignani, C. Corsi, G. Gervasi, C. Leoprini; R. Nieri, A. M. Rossi, G. Stretti and G. Turchi, Evaluation of the genetic effects induced by vinyl chloride monomer (VCM) under mammalian metabolic activation: in vitro and in vivo studies, Mutation Res., **40** (1976) 85–96.
[19] Huberman, E., H. Bartsch and L. Sachs, Mutation induction in Chinese hamster V 79 cells by two vinyl chloride metabolites, chloroethylene oxide and 2-chloroacetaldehyde, Int. J. Cancer, **15** (1975) 539–544.
[20] Vogel, E. and F. H. Sobels, The function of *Drosophila* in genetic toxicology testing, Mutation Res.,
[21] Loprieno, N., R. Barale, S. Baroncelli, H. Bartsch, G. Bronzetti, A. Cammellini, C. Crosi, D. Frezza, R. Nieri, C. Leoprini, D. Rosellini, and A. M. Rossi, Induction of gene mutations and gene conversion by vinyl chloride metabolites in yeast, Cancer Res., (1976) in press.
[22] Purchase, I. F. H., C. R. Richardson and D. Anderson, Chromosomal and dominant lethal effects of vinyl chloride, The Lancet, **II** (1975) 410–411.
[23] Ducatman, V., K. Hirschhorn and I. J. Selikoff, Vinyl chloride exposure and human chromosome aberrations, Mutation Res., **31** (1975) 163–169.
[24] Funes-Cravioto, F., B. Lambert, J. Lindsten, L. Ehrenberg, A. T. Natarajan and S. Osterman-Golkar, Chromosome aberrations in workers exposed to vinyl chloride, Lancet, **1** (1975) 459.
[25] Barbin, A., H. Brésil, A. Croisy, P. Jacquignon, C. Malaveille, R. Montesano and H. Bartsch, Liver-microsome-mediated formation of alkylation agents from vinyl bromide and vinyl chloride, Biochem. Biophys. Res. Commun., **67** (1975) 596–603.
[26] Göthe, R., C. J. Callerman, L. Ehrenberg and C. A. Wachmeister, Trapping with 3,4-dichlorobenzenethiol of reactive metabolites formed in vitro from the carcinogen vinyl chloride, AMBIO, **3** (1974) 234–236.
[27] Grigorescu, I. and G. Toba, Clorura di vinil. Aspecte de toxicologie industriala, Rev. Chim. Rom., **17** (1966) 499–501.
[28] Hefner, R. E., P. G. Watanabe and P. G. Gehring, Preliminary studies of the fate of inhaled vinyl chloride monomer (VCM) in rats, Ann. N.Y. Acad. Sci., **246** (1975) 135–148.
[29] Ames, B. N., W. E. Durston, E. Yamasaki and F. D. Lee, Carcinogens are mutagens: a simple test system combining liver homogenates for activation and bacteria for detection, Proc. Nat. Acad. Sci. (US) **8** (1973) 2281–2285.
[30] Stich, H. F. and R. H. C. San, DNA repair and chromatid anomalies in mammalian cells exposed to 4-nitroquinoline 1-oxide, Mutation Res., **10** (1970) 389–404.

[31] Bartsch, H., C. Malaveille and R. Montesano, Differential effect of phenobarbitone, pregnenolone-16α-carbonitrile and aminoacetonitrile on dialkylnitrosamine metabolism and mutagenicity in vitro, Chem.-Biol. Interactions, **10** (1975) 377–382.
[32] IARC Monographs on the Evaluation of Carcinogenic Risk of Chemicals to Man, Some aromatic amines, hydrazine and related substances, N-nitroso compounds and miscellaneous alkylating agents, International Agency for Research on Cancer, Vol. **4** (1974) 321–328.
[33] Nebert, D. W. and H. V. Gelboin, Substrate-inducible microsomal aryl hydroxylase in mammalian cell culture. I. Assay and properties of induced enzyme, J. Biol. Chem., **243** (1968) 6250–6261.

Monitoring of Human Populations in Connection with Environmental Pollution by the Evaluation of Chromosome Anomalies

N. P. BOCHKOV

Institute of Medical Genetics, USSR Academy of Medical Sciences, Moscow, USSR

The monitoring of the mutation process in man helps to solve different theoretical and applied problems of human genetics, for instance, such as the intensity and direction of hereditary variability in human populations, the stability of genetic structures, the prognosis of the number of patients with genetic diseases, and so on. On the basis of the goals of the symposium, in this report only one aspect of the monitoring of mutation process in man will be considered, namely, the use of this information for evaluation of mutagenic effects of the new factors of the environment. In other words, this report deals with the evaluation of the intensity of the mutation process in the course of time. For the technically grounded approach to the genetic monitoring of human populations in connection with environmental pollution it seems necessary to know all the primary laws of the spontaneous mutation process: the average rate of different types of mutation, the range of fluctuations and their dependence upon different factors, the most adequate methods to study it.

From the technique point of view it is very difficult to study the dynamics of the mutation process in man because the change in the number of mutants in populations may take place not only at the account of the mutation process, but at the account of other genetic processes in populations (genetic drift, selection, migration) as well. This is the reason of the fact that the question of human genetic monitoring of populations is still at the stage of wide discussions, but not fulfilment (Committee 17, 1975; NEEL, 1974; NEEL, BLOOM, 1969; SUTTON, 1970; VOGEL, 1970). Many practical proposals concerning the monitoring of the intensity of gene mutation process in man were not fulfiled because of low rates of gene mutations and significant genetic heterogeneity of phenotypically similarly manifestated mutations. Registration of the human mutation load only for chromosome anomalies is considered unilateral. As it was correctly emphasized by Prof. J. NEEL (1974), the future of monitoring of human populations is with the combination of cytogenetic and biochemical approaches. We shall consider only cytogenetic approaches to monitoring of the mutation process in man, on the basis of modern cytogenetic requirements.

The monitoring of human populations quite evidently must include the evaluation of chromosome and genome mutations in germ and somatic cells (BOCHKOV, 1971). At any rate it is necessary to take into account the following three general principles of study of the mutation process: 1) the preciseness of identification of mutation; 2) the differentiation between the mutations inherited from the previous generations and newly appeared ones; 3) the estimation of the sample size which would permit to evaluate the intensity of the mutation process or the degree of its changes the limits of which are determined in advance.

Monitoring of any type of mutation has its own specific features. To study chromosomal anomalies in populations is sometimes more difficult as compared with gene mutations, and sometimes easier.

One of the most important characteristic feature of chromosomal anomalies is their ability to cause frequently intrauterine and early postnatal deaths. Taking into account this fact, the estimation of the frequency of mutants cannot be carried out by way of the investigation of the whole population or even by way of total examination of newborns. To estimate the mutation rate it is important to study the data on spontaneous abortions and stillbirths. The mutation rate should necessarily be determined on the basis of its frequency in spontaneous abortions, stillbirths, and newborns.

A vast store of information is already available concerning the evaluation of the frequencies of chromosomal anomalies among different groups of examined people (HAMERTON, 1971). Not all of it, however, may be used for monitoring, because only those data may be used for that which were obtained by standard method approaches.

The frequencies of chromosomal anomalies appearing in germ cells cannot be fully determined with modern methods because the death at the earliest stages of the development is not detected. That should necessarily be taken into consideration in general program of monitoring. At cytogenetic investigation alongside with diagnostics of mutations which appeared in germ cells, mutations appearing at early stages of the development in somatic cells, that are cases of mosaicism, are also registered. Though they belong to different types of mutations it is not important from the viewpoint of monitoring of the mutation process in human populations in connection with the environmental pollution, in what kind of cells the mutation appears. It is important that it has a similar pathologic expression and the same causes of appearance. For more special cases of analysis mosaicism should be separated from the general group of mutants.

To discuss the program of monitoring of human populations by chromosomal anomalies first of all it is necessary to consider the frequencies of chromosomal and genome mutations in different investigated groups: among spontaneous abortions, stillborns (and other cases of perinatal death), premature babies, newborns, infants with congenital malformations. The monitoring of the mutation process in somatic cells by registering chromosomal aberrations may be considered as a separate part of the problem (BOCHKOV et al., 1974).

The frequency of chromosomal anomalies in spontaneous abortions

According to statistical data of different countries summarized by WHO, about 15% of pregnancies result in spontaneous abortions (the general number of pregnancies does not contain the cases when a pregnancy results in induced abortions). The frequency of spontaneous abortions ranges from 10 to 20% according to the data by different authors; this may be explained by different registering methods, and different environment in investigated groups. The major part of embryos' deaths occur by the first 12 weeks of pregnancy.

The cytogenetic investigation was carried out among over 4 thousand spontaneous abortions (tissues of embryos, amnion and chorion). Not all the reports deal with the data concerning the nature of anomalies of karyotype, therefore in table 1 there is

Table 1

Results of cytogenetic investigation of spontaneous abortions

Authors, year	Number of observations	Karyotypical anomalies, % aneuploidies	Polyploidies	structural	mosaicism	total
THIEDE, METCALFE, 1966	54	37.0	7.4	0.0	0.0	44.4
CARR, 1967	227	17.2	4.8	0.0	0.0	22.0
BOWEN, LEE, 1969	41	17.1	7.3	2.4	2.4	29.3
ARAKAKI, WAXMAN, 1970	127	29.1	6.3	0.8	9.4	49.6
DHADIAL et al., 1970	423	17.0	4.0	0.0	1.2	23.9
MIKAMO, 1970	67	16.4	7.5	1.5	0.0	*25.4
GEISLER et al., 1971	91	33.0	1.1	0.0	1.1	38.5
LEVY et al., 1972	95	27.4	17.9	2.1	1.1	48.4
KAJII et al., 1973	152	40.8	9.9	2.0	0.7	53.9
THERKELSEN et al., 1974	290	35.9	8.3	0.0	0.0	47.6
BOUE, BOUE, 1975	1498	42.4	16.0	2.3	0.7	61.5
KULIEV, 1972	209	16.3	4.8	0.5	0.0	22.0
Total	3274	32.8	10.8	1.3	0.9	46.9

* Some uncertain cases and double anomalies are included additionally in this column.

presented the material on 3274 abortions with such data. On the whole the range in data by different authors is even higher than those given in table 1: from 8% (STENCHEVER et al., 1967) to over 60% (SZULMAN, 1965; BOUE et al., 1974). First of all these variations in frequencies may be explained by the way of selection of the material on spontaneous abortions for cytogenetic investigation. Only in a few works the sample was total (BOUE a. BOUE, 1975), and in the rest of them it was selected. The first thing is that the incompleteness of the sample is the result of the wrong proportion of the investigated cases according to gestation period. At the same time in a number of works the elimination is shown to occur with more intensity at early stages.

BOUE and BOUE (1975) analyzed the distribution according to gestation period in 1205 spontaneous abortions, which were studied cytogenetically. The information obtained by them showed the following frequencies of chromosomal anomalies: the 2nd week – 78%, the 3rd week – 69%, the 4th week – 61%, the 5th week – 62%, the 6th – 70%, the 7th – 48%, the 8th—12th weeks – 23%. In abortuses of the gestation period from 2 to 7 weeks the frequency of chromosome and genome mutations was on the average, 66%. The similar results were obtained in some other investigations. For instance, THIEDE and METCALFE (1966) found embryos of 4 weeks to have 88% of chromosomal anomalies. SZULMAN (1965) found 64% of anomalies in 8 weeks' abortions, and MIKAMO (1970) – 75% of chromosomal anomalies among abortuses before 5 weeks;

50% – in the group of 5–8 weeks, and 5% – in the group of 9–12 weeks. The last group is very small.

The conclusion of the higher frequency of chromosomal anomalies in spontaneous abortions at early stages of pregnancy is proved by some indirect data: morphologically anomalous embryos can more often be found in abortions at early stages of pregnancy (MILLER, POLAND, 1970), and among them the frequency of chromosomal anomalies is especially high - about 95–98% (MIKAMO, 1970).

Now it is not possible yet to determine exactly the average value (and it is exactly this value that is needed) of the frequency of chromosomal anomalies in spontaneous abortions, taking into account the relative part of abortions at different stages. An approximate analysis shows, however, that the average for the whole period will range between 30% and 50%.

In fact, almost all the cases of chromosomal anomalies are newly appeared. Only half of structural rearrangements are inherited. However, they make only 1.3% of the total number of anomalies.

Consequently, from the point of view of the monitoring of the mutation process, registering of chromosomal anomalies in spontaneous abortions may be equalled to registering of mutants with newly appeared mutations.

The frequency of chromosomal anomalies in perinatal mortality

Perinatal mortality comprises antenatal (before the delivery), intranatal (during the delivery), and postnatal death (during the first 7 days). Perinatal death makes 1–4%, on the average, 2%, in the countries with well-developed health service. Its level ranges significantly depending on the conditions of medical service.

There are many technique difficulties in cytogenetic investigation of the causes of perinatal mortality (the difficulties in obtaining the material, unsatisfactory growth of cultures). That is, evidently, the reason of the fact that only three investigations have been carried out concerning this problem. The results of these investigations are presented in table 2.

Table 2

Frequency of chromosomal anomalies at perinatal mortality

Author, year	Period of death						Total for perinatal period	
	antenatal		intranatal		postnatal			
	investigated	anomalies of kariotype %	investigated	anomalies of kariotype %	investigated	anomalies of kariotype %		
MACHIN et al., 1974	34	8.8	122	4.1	344	5.8	500	5.6
SUTHERLAND et al., 1974	3	33.3	32	3.1	98	7.0	133	6.8
KULESHOV, 1976	22	13.6	67	6.0	113	7.1	202	7.4
Total –	59	11.9	221	4.5	555	6.3	835	6.2

Out of all the anomalies aneuploidy makes 77 %, structural rearrangements are 19 %, and polyploidy – 4 %. As compared with spontaneous abortions at perinatal mortality there is a greater number of structural rearrangements, but mosaicism is never found. Aneuploidy makes the main group which may be considered as newly appeared mutations.

On the basis of the data available in literature it is difficult to draw a conclusion if there is any difference between the frequency of chromosomal anomalies in different periods of ontogenesis. Only the fact of the increased frequency of chromosomal anomalies does not raise any doubts.

The frequency of chromosomal anomalies in children with congenital malformations

As chromosomal anomalies cause severe malformations, it might be considered that chromosomal anomalies occur more often in children with congenital malformations, than in the general group of children. Many chromosomal diseases (EDWARD's, PATAU's, TURNER's, DOWN's, cat cry syndromes) manifest themselves as specific combinations of congenital malformations, which permit to make a diagnose. For instance, the frequency of DOWN's syndrome, determined with clinical methods corresponds to that found at cytogenetic investigations.

Cytogenetic studies on children with congenital malformations are few. This may be explained by the fact that such children make only 3–4 %, and it is difficult for one investigator to collect a big group. In the available material the groups are divided in the following way: 1) with any congenital malformation; and 2) with severe and multiple congenital malformations. These data are presented in table 3.

Table 3

Frequency of chromosomal anomalies among newborns with congenital malformations

Author, year	Any congenital malformations		Multiple congenital malformations	
	investigated	anomalies of karyotype, %	investigated	anomalies of karyotype, %
STEWART et al., 1969	184	4.3	–	–
CHEBOTAREV et al., 1973	1029	5.9	–	–
KORAC et al., 1973	–	–	261	12.6
BOCHKOV et al., 1974	–	–	66	13.6
Total –	1213	5.7	327	12.8

As it is seen in the table, among children with multiple congenital malformations, on the average, 13 % have chromosomal anomalies, and in the general group – about 6 %. It is necessary to note that the general number of children with congenital malformations is several times as much higher than the frequency of children with multiple or severe malformations. This fact may restrict a cytogenetic study on this group of children to monitor the level of the mutation process.

Types of chromosomal anomalies in the group of children with congenital malformations are trisomies 13, 18, 21, and unballanced structural rearrangements.

The frequency of chromosomal anomalies in newborns

The major part of investigations was carried out in groups of newborns. At first the frequency of chromosomal diseases was estimated with clinical methods (trisomy 21) or with the investigation of sex chromatin (anomalies of sex chromosomes). Then with the development of the method of leucocyte culture they gave place to karyotype investigations of newborns without selection. That allowed to estimate the frequency of mutants and types of mutations.

Populations-cytogenetic study on newborns is carried out in 7 laboratories of different countries. The summarized results of these studies are given in table 4.

Table 4

Frequency of chromosomal anomalies among newborns

Author, year	Investigated	Anomalies of sex chromosomes, %	Autosome trisomies, %	Structural rearrangements, %	Total (anomalies), %
Hamerton et al., 1975	13939	0.14	0.13	0.19	0.46
Jacobs et al., 1974	11680	0.27	0.17	0.23	0.67
Nielsen, Sillesen, 1975	11148	0.26	0.16	0.41	0.83
Walzer, Gerald, 1972	9048	0.19	0.08	0.20	0.46
Lubs, Ruddle, 1970	4353	0.25	0.11	0.14	0.51
Sergovich et al., 1969	2081	0.24	0.10	0.14	0.48
Kuleshov, 1976	5766	0.24	0.19	0.31	0.75
Total –	58015	0.22	0.14	0.25	0.61

The frequency of chromosomal anomalies in newborns is 0.61 %, its distribution in the main groups being as follows: anomalies in the system of sex chromosomes is 0.22 % (36.1 % of the general number of anomalies), trisomies of autosomes – 0.14 % (23.0 % of the general number), structural rearrangements – 0.25 % (40.9 % of the general number). Though cytogenetic data permit to carry out the complete analysis of types of anomalies, for the genetic monitoring it is important to have averaged values which may be used in calculations or in planning programs.

In chromosome pathology of newborns there are many anomalies in sex chromosomes (about 2 per 1000 newborns), and trisomy 21 (1.2 per 1000 newborns). Exactly these values are most important for monitoring because the anomalies in sex chromosome are diagnosed with the method of sex chromatin, and Down's syndrome is diagnosed by physicians.

A significant part of the general number of chromosomal anomalies (40.9 %) is structural rearrangements. But they are not fit enough for monitoring because 78 % of them are not newly appeared, but family cases. And it is quite understandable, as the greater part of rearrangements are the ballanced ones.

The approximate calculation of mutation rate

The precise calculation of mutation rate in germ cells cannot be carried out as there are no methods which permit to estimate mutations in gametes and zygotes. The nearest value of the frequencies of the primary events is the registering of all the cases of chromosomal anomalies, no matter what the result of pregnancy is.

The calculation of mutation rate can be made on the basis of the direct method. In fact, all cases of polyploidy, aneuploidy, mosaicism are newly appeared. That makes the calculations easy. Among structural rearrangements sporadic cases make 50% in spontaneous abortions and stillbirths, and 22% in newborns.

For the calculation of mutation rate the following figures may be taken from the data given above: spontaneous abortions – 46.7%, stillborns – 6.2%, newborns – 0.6%. Besides that, it is necessary to allow for family cases. The generalized materials on calculations of the frequency of mutations are given in table 5. Only structural rearrangements may be considered as family cases (50% in spontaneous abortions and stillborns, and 78% in newborns). The number of rearrangements out of the general number of anomalies is given above.

As it is seen in table 5, in 7.4% of pregnancies there are chromosomal anomalies, the greater part of which are eliminated by way of spontaneous abortions.

Table 5

Calculation of the number of chromosome and genome mutations per 1 million of pregnancies

Results of pregnancies	Number of pregnancies		Number of sporadic cases
	Total	With chromosomal anomalies	
Spontaneous abortions	150000	70050	69170
Still-borns	20000	1245	1125
Newborns	830000	5065	3416
Total –	1000000	76360	73711

Principles of genetic monitoring of human populations on the basis of cytogenetic investigation of results of pregnancies

It is accepted among geneticists that the frequency of mutations should be counted per gamete per generation. These figures are used for prognosis of genetic effects of the environmental factors. From the point of view of monitoring of populations it is important to take into account the primary events and to give their comparative estimation (in the course of time; between different groups, and so on).

Monitoring of populations on the basis of registering chromosomal anomalies in progeny is quite reasonable from the method point of view, firstly, because cytogenetically the diagnoses are made correctly; secondly, an inherited defect occurs with high frequency and complete manifestation. There is a small number of inherited cases among them.

The possibilities of monitoring are determined with the exactness of answer given in advance in certain fluctuations of the intensity of the mutation process (for instance, 10%; 50%, or 200%). Confidence limits of frequencies of chromosomal anomalies in different groups are given in table 6. They are based on experimental data summarized in tables 1—4. The frequency of inherited cases is not high, so it may not be taken into consideration.

Table 6

Confidence limits of frequency of chromosomal anomalies (at 95% level of probability)

Investigated group	Increase of mutational level, %	Mean	Limits of fluctuations upper	low	Sample size
Spontaneous abortions	Control	0.467	0.484	0.450	3274
	20	0.560	0.630	0.488	200
	50	0.699	0.875	0.488	25
	100	0.933	0.993	0.578	10
Still-borns	Control	0.062	0.080	0.046	835
	50	0.093	0.107	0.080	1900
	100	0.125	0.177	0.080	200
Newborns with congenital malformations	Control	0.057	0.071	0.044	1213
	50	0.085	0.100	0.071	1500
	100	0.114	0.164	0.072	200
Newborns (total)	Control	0.0061	0.0067	0.0054	58015
	20	0.0073	0.0078	0.0067	90000
	50	0.0091	0.0118	0.0067	5700
	100	0.0121	0.0190	0.0068	1400

The given calculations, however, cannot always be used since, beside chromosomal anomalies, there are other causes of spontaneous abortions, stillbirths, and congenital malformations.

From the monitoring point of view these figures may be applied most adequately in examination of newborns. At the monitoring ot the study on spontaneous abortions first of all it is necessary to determine their frequency and to test the proportion of aberrant and nonaberrant karyotypes. The frequency of spontaneous abortions may serve the main sign of the change of mutation process, therefore the correct technique in their registration is very important. If the rate of the mutation process changes that will cause the increase of deaths of embryos and foetuses, that is the frequency of spontaneous abortions and stillbirths will rise. In this case the relative frequency of embryos with aberrant karyotypes will also rise.

For monitoring of the populations by way of the investigation of newborns, it is not necessary to examine the group totally. It would be more economical to arrange the investigation with determining of the frequency of congenital malformations (with cytogenetic investigation of a selected group of children with congenital malformations), the frequency of sex chromosome anomalies by examination of sex chromatin, and the frequency of DOWN's syndrome clinically.

Due to this information it is possible to get quite enough knowledge of chromosomal pathology on the whole.

The control over the results of a certain number of pregnancies and their cytogenetic study might be the most correct estimation. In table 7 there are given confidence limits (95% level of probability) of frequencies of chromosomal anomalies in all results of pregnancies (15% — spontaneous abortions, 2% — stillborns, 83% — newborns). The analysis shows that on condition of 10% increase of the frequency of chromosome and genome mutations it is enough to investigate cytogenetically 6050 results of pregnancies (about 900 spontaneous abortions, 120 stillbirths and over 5000 newborns), and at the double increase to study only 85 results of pregnancies.

Table 7

Confidence limits of frequency of chromosomal anomalies at total cytogenetic investigation of results of pregnancy (at 95% level of probability)

Increase of mutational level, %	Mean frequency	Limits of fluctuations		Sample size
		upper	low	
Control	0.0730	0.0735	0.0725	1000000
100	0.1461	0.2356	0.0751	85
50	0.1096	0.1516	0.0736	270
20	0.0877	0.1026	0.0738	1550
10	0.0804	0.0874	0.0736	6050

Monitoring of human populations may be carried out in the general group without an identification of some factor or in a special group of people (for instance, those living in the zone of pollution or working at some plants). On this basis the monitoring schemes may be recommended.

In the first case the representative sample of the general population and subpopulations is studied. The comparison of mutational levels can be made in periods of time or between different zones. From the method point of view such monitoring should correspond to population-cytogenetic study of different results of pregnancies. As it was shown above two variants are most suitable in this case: 1) estimation of frequencies of spontaneous abortions with partial cytogenetic investigations; 2) estimation of frequencies of chromosomal diseases among newborns by determining the number of children with congenital malformations (and their cytogenetic examination), with Down's syndrome and sex chromosome anomalies by sex chromatin.

In the second case it is necessary to study the group limited in number which is exposed to some factor. As a rule, these groups consist of several thousands of people, and rarely more. Thus, it is reasonable for them to analyse all results of pregnancies and determine the sample size according to table 7.

Monitoring of human populations by study of chromosomal aberrations in somatic cells

The estimation of genetic effects of the environmental factors in human somatic cells may be a good contribution to the monitoring of the effects in progeny. With this purpose it is possible to study chromosomal aberrations in a lymphocyte culture for which in many laboratories the spontaneous level has already been established.

As it is known, the exactness of estimation of some value is determined by its error which depends upon the type of distribution of measured individual values, parameters of this distribution and the sample size.

In table 8 there are given the data of the Laboratory of Mutagenesis of the Institute of Medical Genetics on the study of the frequency of spontaneous chromosomal aberrations in the culture of lymphocytes in healthy people of both sex and different age (the total number of individuals is 437). These individuals were not exposed to radiation, they did not take any chemicals, and were not ill with virus diseases. In cytogenetic investigation only aberrations of all types were registered, but "gaps" were not considered aberrations.

Table 8

Frequency of spontaneous chromosomal aberrations in lymphocyte cultures of individuals of different age and sex

Age group	Sex	Number of individuals	Investigated cells	Cells with aberrations, %	Number of aberrations per 100 cells
newborns	m	58	6641	1.29	1.33
	f	49	5548	1.19	1.19
	m + f	107	12189	1.25	1.26
20–30 years	m	153	22776	1.20	1.27
	f	104	16255	1.11	1.13
	m + f	257	39031	1.16	1.21
50–70 years	m	51	5800	1.34	1.41
	f	22	3000	1.03	1.03
	m + f	73	8800	1.24	1.28
Total	m	262	35217	1.24	1.30
	f	175	24803	1.12	1.14
	m + f	437	60020	1.19	1.23

As it is seen in table 8 the frequency of spontaneous chromosome aberrations in individuals of different age and sex does not differ. The mean number of cells with aberrations is 1.19% (1.24 aberrations per 100 cells). This means that the cells with two aberrations occur very seldom.

For monitoring of population it is very important to know the type of distribution of cultures according to the frequency of spontaneous aberrations. These data are given in table 9. In each culture 100 (sometimes 200) cells were investigated. Check-up of the correspondence between the empiric distribution of cultures and Poisson's and binomial distributions did not show any statistical differences between them. The mean value of the frequency of cells with chromosome aberrations for these distributions is 1.18%. On the basis of the empiric and theoretical data it may be seen that the maximum spontaneous level of aberrant metaphases in lymphocyte cultures of human peripheral blood very seldom exceeds 3.0%.

Table 9

Distribution of human lymphocyte cultures according to frequency spontaneous chromosomal aberration

Cells with aberrations, %	Number of cultures	Theoretical distribution	
		poissons	binomial
0	160	161.9	160.8
1	201	191.1	192.0
2	105	112.7	113.5
3	44	44.3	44.3
4	11	13.1	12.8
5	6	3.9	3.6
Total	527	527.0	527.0

For monitoring of human populations by chromosomal aberrations in somatic cells it is necessary to calculate sample sizes (the number of individuals and the number of investigated cells). In table 10 there are the corresponding data in different estimations of changes of the mutation process. The calculations are based of the spontaneous level of cells with chromosomal aberrations (M = 1.19%) determined in the Laboratory

Table 10

Sample size for registration of mutagenic effects of different level

Increase of mutation level, %	Size of population	Sample size representing the population	
		number of individuals	number of cells
20	500	288	28800
	10000	357	35700
	100000	369	36900
50	500	62	6200
	10000	70	7000
	100000	70	7000
100	500	20	2000
	10000	20	2000
	100000	20	2000

of Mutagenesis, and registering of the effects with 95% probability. As it is seen in table 10, the increase of spontaneous level of chromosomal aberrations is not difficult to establish even in one laboratory, especially if the double increase is taken (it is enough to investigate 2000 cells). In those laboratories or for those populations in which the spontaneous level is not known, the work should be started with its estimation

Thus, in conclusion it may be said that cytogenetic methods can be used quite well for genetic monitoring of human populations. The information obtained by cytogeneticists permit to exactly determine the technique requirements and sample size necessary for registration of the changes in the mutation process the given limits.

References

ARAKAKI, D. T. and S. H. WAXMAN, Effect of gestational and maternal age in early abortion. Obstet. and Ginecol., **35** (1970) 264–269.

BOCHKOV, N. P., 1971, Human chromosomes and radiation, m. Atomizdat.

BOCHKOV, N. P., N. P. KULESHOV, A. N. CHEBOTAREV, V. I. ALEKHIN, and S. A. MIDIAN, Population cytogenetic investigation of newborns in Moscow, Humangenetik, **22**, 2 (1974) 139–152.

BOCHKOV, N. P., R. J. ŠRÁM, N. P. KULESHOV and V. S. ZHURKOV, System for the evaluation of the risk from chemical mutagens for man: basic principles and practical recomendations, Mut. Res., **38** (1976) 191–202.

BOUE, A. and J. BOUE, 1975, Chromosome abnormalities and abortion. In: Phisiology and genetics of reproduction, pt. B (Coutinho a. Fucks eds.), Plenum Publ., N.Y., p. 317–339.

BOUE, I., A. BOUE and P. LAZAR, The epidemiology of human spontaneous abortions with chromosomal anomalies. In: The Biology and Pathology of aging gametos (R. I. Blandau, ed.), Karger, Basel, 330–348.

BOWEN, P. and C. S. N. LEE, Spontaneous abortion, Chromosome studies on 41 cases and an analysis of maternal age and duration of pregnancy in relation to karyotype, Amer. J. Obstet. Gynecol., **104** (1969) 973–983.

CARR, D. H., Chromosome anomalies as a cause of spontaneous abortion, Amer. J. Obstet. and Gynecol., **97** (1967) 283–293.

CHEBOTARĖV, A. N., N. P. BOCHKOV and O. P. KIRICHENKO, Population-cytogenetical studies of newborns with congenital malformations, Genetics (USSR), **9**, 2 (1973) 146–150.

Committee 17, Environmental mutagenic hazards, Mutagenicity screening in now both feasible and necessary for chemicals entering the environment., Science, **187** (1975) 503–514.

DHADIAL, R. K., A. M. MACHIN and S. M. TAIT, Chromosomal anomalies in spontaneously.aborted fetuses, Lancet, **11** (1970) 20–21.

GEISLER, M., J. KLEINEBRECHT und K.-H. DEGENHARDT, Histologische Analyse von triploïden Spontanaborten, Humangenetik, **16**, 4 (1972) 283–294.

JACOBS, P. A., M. MELVILLE, S. RATELIFFE, A. KEAY and Y. SYME, A cytogenetic survey of 11680 newborn infants, Ann. Hum. Genet., London, **37** (1974) 359–376.

HAMERTON, J. L., 1971, Human Cytogenetics, vol. 1,2, New York, Academic Press.

HAMERTON, J. L., M. RAY, J. ABBOT, C. WILIAMSON and G. S. DUCASSE, Chromosome studies in a neonatal population, Canad. Med. Ass. J., **106** (1972) 776–779.

KAJII, T., K. OHAMA, N. NIIKAWA, A. FERRIER and S. AVIZACHAN, Banding analysis of abnormal karyotypes in spontaneous abortion, Amer. J. Hum. Genet., **25**, 9 (1973) 539–547.

KORAC DANICA, B. GARZICIC and B. AJDARIC, Cytogenetika ispitivanja u odojcadi sa Kogenitalnim anomalijama, Acta biol iugosl., **5**, 3 (1973) 345.

KULESHOV, N. P., Chromosome anomalies of infants dying during the perinatal period and premature newborn, Humangenetik, **31**, 2 (1976) 151–160.

KULESHOV, N. P., 1976 (Personal communications).

KULIEV, A. M., Correlation between phenotype and karyotype anomalies in embryogenesis, Genetics (USSR), **8**, 7 (1972) 140–153.

LEVY, J., P.-A. CHADEYRON and B. FONTY, 1972, Abortements spontanes et triploidie, Rev. franc. Gynecol., **67**, 5 (1972) 327–342.
LUBS, H. A. and F. N. RUDDLE, Chromosome abnormalities in the human population estimation of rates based on New Haven newborn study, Science, **169** (1970) 495–497.
MACHIN, G. A. and J. A. CROLLA, Chromosome constitution of 500 infants dying during the perinatal period, Humangenetik, **23** (1974) 183–198.
MIKAMO, K., Anatomic and chromosome anomalies in spontaneous abortion, Amer. J. Obstet. and Gynecol., **106** (1970) 243–254.
MILLER, J. R. and B. J. POLAND, The value of human abortuses in the surveillance of development anomalies, I. General overview, Canad. Med. Assoc. J., **103** (1970) 501–507.
NEEL, J. V., Developments in monitoring human population for mutation rates, Mut. Res., **26** (1974) 319–328.
NEEL, J. W. and A. D. BLOOM, The detection of environmental mutagens, Med. Clinics of N. Amer. **53**, (1969) 1243–1256.
NIELSEN, J. and I. SILLESEN, Incidence of chromosome aberrations among 11148 newborn children, Humangenetik, **30** (1975) 1–12.
SERGOVICH, F., G. H. VALENTINE, A. T. L. CHEN, R. A. H. KINCH and M. S. SMOUT, Chromosome aberrations in 2159 consecutive newborn babies, New Engl. J. Med., **280** (1969) 851–855.
STENCHEVER, M. A., J. A. HEMPEL and M. N. MACINTYRE, Cytogenetics of spontaneously aborted human fetuses, Obstet. Gynecol., **30** (1967) 683–691.
STEWARG, A. L., A. J. KEAY, P. A. JACOBS and M. M. MELVILLE, A chromosome survey of unselected liveborn children with congenital abnormalities, J. Paediatr., **74**, 3 (1969) 449–458.
SUTHERLAND, G. R., R. BAULD and A. D. BAIN, Chromosome abnormality and perinatal death, Lancet, 7860, (1974) 752.
SUTTON, H. E., Workshop on Monitoring of Human Mutagenesis, Teratology, **4**, 1 (1971) 103–107.
SZULMAN, A. E., Chromosomal aberrations in spontaneous human abortions, New Engl., J. Med., **272** (1965) 811–818.
THERKELSEN, A. J., N. GRUNNET, T. HJORT, M. JENSEN, I. JONASSON, J. G. LAURITZON, J. LINDSTEN and B. G. PETERSEN, 1974, Studies on spontaneous abortion, In: Symp. on Chromosome Errors in relation to reproductive failure, Paris (A. Boue and C. Thibault, eds) Inserm, Paris, 81.
VOGEL, F., 1970, Monitoring of human populations, In: Chemical Mutagenesis in mammals and man. (Vogel, F. and Röhrborn eds), Berlin, Springer, 445.
THIEDE, H. A. and S. METCALFE, 1966, Chromosomes and human pregnancy wastage, Amer. J. Obstet. and Gynecol., **96** (1966) 1132–1138.
WALZER, S. and P. S. GERALD, Chromosome abnormalities in 11154 newborn infants, Amer. J. Hum. Genet., **24** (1972) 38a.

Authors

Z. Bardoděj, Institute of Hygiene and Epidemiology, Šrobarova 48, 10042 Prague 10 (ČSSR).
I. Barrai, Universitá di Ferrara, Instituto di Zoologia e Biologia Generale, Via L. Borsari 46, 44100 Ferrara (Italy).
H. Bartsch, Unit of Chemical Carcinogenesis, International Agency for Research on Cancer, 150 cours Albert Thomas, 69008 Lyon (France).
A. Basler, Institut für Humangenetik und Anthropologie, Universität Düsseldorf, 4000 Düsseldorf (FRG).
N. P. Bochkov, Institute of Medical Genetics, Academy of Medical Sciences of the USSR, Kashirskoye shosse 6A, 115478 Moscow (USSR).
R. C. von Borstel, Department of Genetics, University of Alberta, Edmonton, Alberta T6G 2E9 (Canada).
R. Braun, Zentralinstitut für Genetik und Kulturpflanzenforschung der Akademie der Wissenschaften der DDR, 4325 Gatersleben (GDR).
B. A. Bridges, Medical Research Council, Cell Mutation Unit, University of Sussex, Falmer, Brighton BN1 9QG (UK).
C. H. Clarke, School of Biological Sciences, U.E.A., Norwich NR4 7TJ (UK).
F. J. de Serres, Mutagenesis Branch, National Institute of Environmental Health Sciences, P.O. Box 12233, Research Triangle Park, NC 27709 (USA).
N. P. Dubinin, Institute of General Genetics, Academy of Science of the USSR, Profsoyuznaya St. 7, 117312 Moscow (USSR).
L. Ehrenberg, Wallenberglaboratoriet, Stockholms Universitet, Lilla Frescati, 10405 Stockholm 50 (Sweden).
S. Hussain, Wallenberglaboratoriet, Stockholms Universitet, Lilla Frescati, 10405 Stockholm 50 (Sweden).
B. J. Kilbey, Institute of Animal Genetics, University of Edinburgh, West Mains Road, Edinburgh EH9 3JN (UK).
W. Klinger, Institut für Pharmakologie und Toxikologie der Friedrich-Schiller-Universität, Zentraler Platz, 69 Jena (GDR).
N. Loprieno, Laboratorio di Mutagenesi e Differenziamento – CNR, c/o Istituto di Genetica dell' Università, Viale Matteotti 1/A, 56100 Pisa (Italy).
P. Oftedal, Institute of General Genetics, University of Oslo, P. O. Box 1031, Blindern, Oslo 3 (Norway).
G. Röhrborn, Institut für Humangenetik und Anthropologie, Universität Düsseldorf, 4000 Düsseldorf (FRG).
J. Schöneich, Zentralinstitut für Genetik und Kulturpflanzenforschung der Akademie der Wissenschaften der DDR, 4325 Gatersleben (GDR).
R. J. Šrám, Institute of Hygiene and Epidemiology, Šrobarova 48, 10042 Prague 10 (ČSSR).

V. S. Zhurkov, Institute of Medical Genetics, Academy of Medical Sciences of the USSR, Kashirskoye shosse 6A, 115478 Moscow (USSR).
P. J. Hastings, Department of Genetics, University of Alberta, Edmonton, Alberta T6G 2E9 (Canada).
Cornelia Schroeder, Institut für medizinische und allgemeine Mikrobiologie des Medizinischen Bereichs der Humboldt-Universität zu Berlin, Lehrstuhl Virologie, DDR 104 Berlin, Schumannstr. 20/21 (GDR).

Bei Fragen zur Produktsicherheit wenden Sie sich bitte an:
If you have any questions regarding product safety,
please contact:

Walter de Gruyter GmbH
Genthiner Straße 13
10785 Berlin
productsafety@degruyterbrill.com